W9-BBP-133

# INCOMPLETE BLOCK DESIGNS

# LECTURE NOTES IN STATISTICS

A Series Edited By

D. B. OWEN

Department of Statistics
Southern Methodist University
Dallas, Texas

Vol. 1 Incomplete Block Designs, *by Peter W. M. John*

*Additional Volumes in Preparation*

# INCOMPLETE BLOCK DESIGNS


## Peter W. M. John

*Department of Mathematics*
*University of Texas at Austin*
*Austin, Texas*



MARCEL DEKKER, Inc. New York and Basel

Library of Congress Cataloging in Publication Data

John, Peter William Meredith.
Incomplete block designs.

(Lecture notes in statistics ; v. 1)
1. Incomplete block designs. I. Title.
II. Series: Lecture notes in statistics (New York, N.Y., 1980- ) ; v. 1.
QA279.J64 001.4'34 80-22866
ISBN 0-8247-6995-3

COPYRIGHT © 1980 by MARCEL DEKKER, INC. ALL RIGHTS RESERVED

Neither this book nor any part may be reproduced or transmitted in any form or by any means, electronic or mechanical, including photocopying, microfilming, and recording, or by any information storage and retrieval system, without permission in writing from the publisher.

MARCEL DEKKER, INC.

270 Madison Avenue, New York, New York 10016

Current printing (last digit):

10 9 8 7 6 5 4 3 2 1

PRINTED IN THE UNITED STATES OF AMERICA

# PREFACE

The topic of incomplete block designs has become attractive to algebraists and combinatorists as well as to statisticians, and contributions are to be found in a bewildering number of journals in several countries. Unfortunately, most courses in the design of experiments never manage to find time for more than a cursory glance at incomplete block designs, and the student does not usually hear more than a few words confined to the intrablock analysis of balanced designs.

In the fall of 1977 I gave a course of lectures at The University of Texas on incomplete block designs. This set of lecture notes is based upon the notes that I prepared for that course. They reflect the fact that my views on the topic had changed since I wrote the chapters on incomplete block designs for my 1971 book. The lectures were not intended to be exhaustive and the bibliography refers only to the works cited in the text. The reader seeking further references will find some of them in my 1971 book or in Raghavarao's book which appeared in the same year.

These lectures focus upon two themes. The first of these is the construction of incomplete block designs by the cyclic development of one or more initial blocks, an idea which was introduced by Bose in his paper on the construction of balanced designs (1939). The second theme comes from the realization by S. C. Pearce that for partially balanced incomplete block designs the concordance matrix and the generalized inverse of the intrablock matrix have the same structure.

My interest in incomplete block designs goes back to 1957-1961, when I had the good fortune to work with the late Henry Scheffé at Berkeley. In August, 1976 I met Professor Pearce at the International Biometrics

Conference in Boston. In the fall of that year I visited Dr. J. A. John and Professor T. F. M. Smith at Southampton University; the influence of their work is apparent in the last chapter.

I owe a particular debt of gratitude to my valiant typist, Suzy Crumley, whose patience and persistence have been admirable.

Peter W. M. John

# CONTENTS

# Chapter One

# INTRODUCTION

1.1 Background. The basic ideas and the jargon of the topic of incomplete block designs come from agricultural experiments, particularly those carried out at the Rothamsted agricultural experiment station in England under the influence of Sir Ronald A. Fisher and Frank Yates. The applications of these ideas are now found in many areas of science and engineering. A typical example of the kind of experiment in which we shall be primarily interested is to compare the yields of several varieties of wheat, barley or some other crop. In a large field the fertility of the ground may vary considerably. The field is therefore divided into relatively small strips called blocks, each of which can be considered as homogeneous in fertility. The blocks are divided into plots. In each plot a single variety is sown, and the yield of the plot is observed. The yields are the data.

A similar experiment would involve comparing fertilizers; we could use the same variety of wheat throughout and apply various fertilizers (or quantities of fertilizers) to the plots. In an industrial experiment the "plots" might be runs on a pilot plant at different temperatures. To embrace all these possibilities we shall talk of experiments to compare treatments, and we shall let $t$ denote the number of treatments. One treatment is applied to each plot. The yields or other observed responses are the data. The term "treatment" in our examples has included varieties, fertilizers and temperatures. In Chapter 12 the "treatments" will be combinations of levels of factors in a multifactor experiment.

The simplest experiment is that in which each block contains exactly one plot with each treatment. This is the complete block experiment. To compare two treatments we have only to compare their average yields over all the blocks. In this monograph we shall be concerned with experiments in which the blocks are not large enough to accommodate all the treatments. In such a case one of the treatments may be so fortunate as to be used in the more fertile blocks, while another may be less well favored. Comparing these treatments by looking at their average yields is no longer reasonable: we must somehow adjust for the block differences.

In this chapter and the next we shall introduce the general incomplete block design and discuss the intrablock analysis, which, in essence, compensates for the differences in fertility between blocks by comparing the yields of plots in the same block, and combining these comparisons. The later chapters will be concerned with the problem of designing incomplete block experiments. How do we decide which treatments are to be used in which block? We shall mean by a design an arrangement of $t$ treatments in $b$ blocks, and we shall discuss the analysis and construction of both balanced and partially balanced designs. In the last chapter we shall discuss the problem of carrying out factorial experiments in incomplete blocks.

There is not enough time available to give an exhaustive discussion of these various topics, and we shall therefore focus our presentation upon two themes. The first of these is found in a paper presented by S. C. Pearce to the Royal Statistical Society in 1963. This may be stated in the simplest case as follows. We associate with each design two matrices. One is the concordance matrix, denoted by $\underset{\sim}{N}\underset{\sim}{N}'$; it is a square matrix of order $t$ which has for the element in the $h^{th}$ row and the $i^{th}$ column the number of blocks in which the $h^{th}$ and $i^{th}$ treatments both appear. The other matrix, $\underset{\sim}{\Omega}$, is equivalent to the variance-covariance matrix for comparing treatment effects. The essence of this theme is that for a broad class of designs $\underset{\sim}{N}\underset{\sim}{N}'$ and $\underset{\sim}{\Omega}$, the one a matrix which summarizes the design, and the other a matrix which summarizes the analysis, have the same pattern. They also have the same latent vectors, and these vectors form an important set of contrasts between treatment effects.

The second theme relates to the pioneer work by R. C. Bose on the construction of incomplete block designs and is the use of cyclic development for obtaining designs. In its basic form we denote the treatments by the integers $0, 1, \ldots, t-1$ and start with a block which is a subset of $k$ treatments. In a cyclic design with $t$ blocks we obtain the $j^{th}$ block $(j = 0, 1, \ldots, t-1)$ by adding $j$ to each element of the initial block, and reducing the integers $\bmod t$. This procedure is called developing the initial block cyclically.

By the same token, no attempt will be made to provide a complete bibliography of the field. References will be made to some books and papers which deal with specific points in the discussion. A good source of references up to 1971 is the book by Raghavarao (1971).

1.2 An example. Suppose that we have four treatments A, B, C and D to be compared in four blocks of three plots each, and that we chose to allocate A, B and C to the first block, A, B and D to the second, A, C and D to the third and B, C, D to the fourth. If we had chosen to develop the initial block 0 1 2 we would have obtained the same design with the blocks in a different order.

We take the following model. Let $y_{ij}$ denote the yield on the plot with the $i^{th}$ treatment in the $j^{th}$ block (if there is one); i takes the values 1, 2, 3, 4 corresponding to A, B, C and D respectively. We then write

$$y_{ij} = \mu + \tau_i + \beta_j + e_{ij}$$

where $\mu$, $\tau_i$, $\beta_j$ are unknown parameters; $\mu$ represents an overall average yield; $\tau_i$ is the contribution of the $i^{th}$ treatment (the $i^{th}$ treatment effect) and $\beta_j$ is the contribution of the $j^{th}$ block (the $j^{th}$ block effect). The last term, $e_{ij}$, is the random component representing the noise in the system. We shall assume that the $e_{ij}$ are independent random variables and are normally distributed each with zero expectation and common variance $\sigma^2$. (The assumption of normality will not be needed to derive the least squares estimates that we shall use; it is enough for them that the error terms be uncorrelated with zero expectations and variance $\sigma^2$).

If we arrange the yields in a table with rows corresponding to blocks the data are

$$\begin{array}{cccc} y_{11} & y_{21} & y_{31} & \\ y_{12} & y_{22} & & y_{42} \\ y_{13} & & y_{33} & y_{43} \\ & y_{24} & y_{34} & y_{44} \, . \end{array}$$

The primary purpose of our experiment is to compare treatments. Suppose that we wish to compare A and B, i.e., to estimate $\tau_1 - \tau_2$.

In the first block $y_{11} - y_{21}$ gives us an unbiased estimate of $\tau_1 - \tau_2$. It is an intrablock contrast, a contrast between two plots in the same block; we mean by an intrablock contrast in this context a linear combination of observations, $\Sigma c_{ij} y_{ij}$, such that $\Sigma_i c_{ij} = 0$ for each j. The block effect $\beta_1$ cancels out. A similar thing happens for $y_{12} - y_{22}$. Is there additional information about $\tau_1 - \tau_2$ in the other blocks in which A and B do not both appear?

Consider the contrast $2y_{13} - y_{33} - y_{43}$ in the third block. It is an unbiased estimate of $2\tau_1 - \tau_3 - \tau_4$. The contrast $2y_{24} - y_{34} - y_{44}$ estimates $2\tau_2 - \tau_3 - \tau_4$. Their difference divided by two gives us a third estimate of $\tau_1 - \tau_2$. How should we combine these three estimates, which are clearly independent, to obtain a "best" composite estimate?

Denote the three estimates by $x_1$, $x_2$, $x_3$ respectively. In forming a linear combination it is obviously reasonable to give $x_1$ and $x_2$ the same weight; what weight shall we give to $x_3$? Consider the linear combination

$$x = (x_1 + x_2 + wx_3)/(2 + w),$$

which is an unbiased estimate of $\tau_1 - \tau_2$. We choose w so as to minimize the variance of x which is given by

$$V(x) = (2 + w)^{-2} \{2 + 2 + (4 + 1 + 1 + 4 + 1 + 1)w^2/4\}$$

$$= (12w^2 + 16)/4(w + 2)^2.$$

V(x) is minimized with $w = 2/3$.

We shall see in the next section that this estimate

$$x = (3x_1 + 3x_2 + 2x_3)/8$$

is the least squares estimate.

1.3 <u>Least squares estimates</u>. Our method of estimating $\tau_1 - \tau_2$ in the last section was clearly ad hoc. We shall now derive the least squares estimates.

We shall denote the least squares estimates of the parameters by $\hat{\mu}$, $\hat{\tau}_i$, $\hat{\beta}_j$. Let $\hat{y}_{ij} = \hat{\mu} + \hat{\tau}_i + \hat{\beta}_j$ and let $d_{ij} = y_{ij} - \hat{y}_{ij}$; $\hat{y}_{ij}$ is the estimated value for the $(ij)^{th}$ plot and $d_{ij}$ is the corresponding residual; $S_e = \Sigma_i \Sigma_j d_{ij}^2$ is the residual sum of squares. We choose $\hat{\mu}$, $\hat{\tau}_i$, $\hat{\beta}_j$ so as to minimize $S_e$.

Differentiating $S_e$ with respect to each of the parameters in turn and equating each derivative to zero gives us the set of normal equations to be solved for the parameters. In this example there are nine such equations. We have $S_e = \Sigma_i \Sigma_j (y_{ij} - \hat{\mu} - \hat{\tau}_i - \hat{\beta}_j)^2$ for all pairs i, j that occur in the data. The following three are typical. There are twelve observations, three on each treatment and three in each block.

$$\partial S_e/\partial\hat{\mu} = 0: \qquad G = 12\hat{\mu} + 3\Sigma\hat{\tau}_i + 3\Sigma\hat{\beta}_j,$$

$$\partial S_e/\partial\hat{\tau}_i = 0: \qquad T_i = 3\hat{\mu} + 3\hat{\tau}_i + \Sigma_j n_{ij}\hat{\beta}_j, \quad i = 1,2,3,4,$$

$$\partial S_e/\partial\hat{\beta}_j = 0: \qquad B_j = 3\hat{\mu} + \Sigma_i n_{ij}\hat{\tau}_i + 3\hat{\beta}_j, \quad j = 1,2,3,4;$$

in these equations $G$ is the grand total of all the observations, $T_i$ is the total of the yields on the $i^{th}$ treatment, $B_j$ is the total for the $j^{th}$ block and $n_{ij}$ is the number of plots in the $j^{th}$ block that contain the $i^{th}$ treatment; $n_{ij} = 0$ or 1.

We proceed by eliminating the estimates of the unwanted parameters $\beta_j$ from the equations.

Let $Q_i = T_i - \Sigma_j n_{ij}B_j/3$; $Q_i$ is called the adjusted treatment total for the $i^{th}$ treatment. The estimates of $\beta_j$ drop out and we have, for example,

$$Q_1 = T_1 - (B_1 + B_2 + B_3)/3$$

$$= 2\hat{\tau}_1 - 2(\hat{\tau}_2 + \hat{\tau}_3 + \hat{\tau}_4)/3;$$

$$Q_2 = T_2 - (B_1 + B_2 + B_4)/3$$

$$= 2\hat{\tau}_2 - 2(\hat{\tau}_1 + \hat{\tau}_3 + \hat{\tau}_4)/3,$$

so that $(\hat{\tau}_1 - \hat{\tau}_2) = 3(Q_1 - Q_2)/8$. If we write this expression in terms of the individual yields, $y_{ij}$, we can see directly that this estimate is the same as the estimate, x, that we obtained by our ad hoc approach in the previous section.

If we were to conclude that there were no block effects, i.e., that $\underset{\sim}{\beta} = 0$, the least squares estimate of $\tau_1 - \tau_2$ would become $(T_1 - T_2)/3 = \bar{y}_1 - \bar{y}_2$, which is the difference between the two treatment means, and not the same as the estimate that we have just obtained. Thus we see that the treatment effects and the block effects are not orthogonal to each other. For this reason Pearce refers to incomplete block designs as nonorthogonal designs.

# Chapter Two

## THE GENERAL INCOMPLETE BLOCK DESIGN

2.1 Introduction. We consider a design for $t$ treatments in $b$ blocks. The $i^{th}$ treatment $(i = 1,2,\ldots,t)$ appears in $r_i$ plots, i.e., has $r_i$ replicates; the $j^{th}$ block contains $k_j$ plots of which $n_{ij}$ receive the $i^{th}$ treatment. We introduce three matrices: $\underset{\sim}{R}$ is the $(t \times t)$ diagonal matrix whose $i^{th}$ diagonal element is $r_i$; $\underset{\sim}{K}$ is the $(b \times b)$ diagonal matrix whose $j^{th}$ diagonal element is $k_j$; the incidence matrix, $\underset{\sim}{N} = (n_{hj})$ has $t$ rows and $b$ columns; $n_{hj}$ is the number of times that the $h^{th}$ treatment appears in the $j^{th}$ block. The incidence matrix completely specifies the design. We shall use $\underset{\sim}{1}_m$ to denote a vector of $m$ unit elements and we shall write $\underset{\sim}{J}_m$ for the square matrix $\underset{\sim}{1}_m\underset{\sim}{1}_m'$; we shall often omit the subscript $m$. It will sometimes be convenient to write $\underset{\sim}{R}\underset{\sim}{1} = \underset{\sim}{r}$, and $\underset{\sim}{K}\underset{\sim}{1} = \underset{\sim}{k}$.

If $n$ is the total number of plots we now have the following relationships between the matrices of the design:

$$\underset{\sim}{R}\underset{\sim}{1} = \underset{\sim}{N}\underset{\sim}{1} = \underset{\sim}{r}\ ,\ \underset{\sim}{K}\underset{\sim}{1} = \underset{\sim}{N}'\underset{\sim}{1} = \underset{\sim}{k},$$

$$\underset{\sim}{1}'\underset{\sim}{R}\underset{\sim}{1} = \underset{\sim}{1}'\underset{\sim}{K}\underset{\sim}{1} = \underset{\sim}{1}'\underset{\sim}{N}\underset{\sim}{1} = n.$$

The model that we introduced for the example in the previous chapter has to be modified slightly. Let $y_{ijm}$ be the yield of the $m^{th}$ of those plots of the $j^{th}$ block which receive the $i^{th}$ treatment, $m = 0,1,\ldots,n_{ij}$:

$$y_{ijm} = \mu + \tau_i + \beta_j + e_{ijm}$$

where the random error terms $e_{ijm}$ are independent normal with $E(e_{ijm}) = 0$, $V(e_{ijm}) = \sigma^2$. If $n_{ij} = 0$ or $1$ for all $i$ and all $j$ we omit the superfluous suffix $m$. The treatment and block effects, $\tau_i$ and $\beta_j$, are regarded as unknown constants and not as random variables.

2.2 <u>The least squares intrablock estimates.</u> We again seek the least squares estimates $\hat{\mu}$, $\hat{\tau}_i$, $\hat{\beta}_j$ of the parameters $\mu$, $\tau_i$, $\beta_j$ by minimizing the residual sum of squares, $S_e$. There are $t+b+1$ normal equations typified by

$$G = n\hat{\mu} + \Sigma_i r_i \hat{\tau}_i + \Sigma_j k_j \hat{\beta}_j,$$

$$T_i = r_i\hat{\mu} + r_i\hat{\tau}_i + \Sigma_j n_{ij}\hat{\beta}_j, \qquad i = 1,\dots,t,$$

$$B_j = k_j\hat{\mu} + \Sigma_i n_{ij}\hat{\tau}_i + k_j\hat{\beta}_j, \qquad j = 1,\dots,b.$$

Turning to matrices we write $\underset{\sim}{T}$ for the vector of treatment totals, $T_i$, $\underset{\sim}{B}$ for the vector of block totals, $B_j$, and $\underset{\sim}{\hat{\tau}}$, $\underset{\sim}{\hat{\beta}}$, for the vectors of estimates of the parameters $\underset{\sim}{\tau}$ and $\underset{\sim}{\beta}$, and obtain

$$\begin{bmatrix} G \\ \underset{\sim}{T} \\ \underset{\sim}{B} \end{bmatrix} = \begin{bmatrix} n & \underset{\sim}{1}'\underset{\sim}{R} & \underset{\sim}{1}'\underset{\sim}{K} \\ \underset{\sim}{R}\underset{\sim}{1} & \underset{\sim}{R} & \underset{\sim}{N} \\ \underset{\sim}{K}\underset{\sim}{1} & \underset{\sim}{N}' & \underset{\sim}{K} \end{bmatrix} \begin{bmatrix} \hat{\mu} \\ \underset{\sim}{\hat{\tau}} \\ \underset{\sim}{\hat{\beta}} \end{bmatrix} .$$

We eliminate the unwanted $\underset{\sim}{\hat{\beta}}$ by multiplying both sides on the left by

$$\begin{bmatrix} 1 & \underset{\sim}{0} & \underset{\sim}{0} \\ \underset{\sim}{0} & \underset{\sim}{I}_t & -\underset{\sim}{N}\underset{\sim}{K}^{-1} \\ \underset{\sim}{0} & -\underset{\sim}{N}'\underset{\sim}{R}^{-1} & \underset{\sim}{I}_b \end{bmatrix},$$

giving three sets of equations:

(i) $$G = n\hat{\mu} + \underset{\sim}{1}'\underset{\sim}{R}\underset{\sim}{\hat{\tau}} + \underset{\sim}{1}'\underset{\sim}{K}\underset{\sim}{\hat{\beta}} ;$$

(ii) $$\underset{\sim}{T} - \underset{\sim}{N}\underset{\sim}{K}^{-1}\underset{\sim}{B} = (\underset{\sim}{R} - \underset{\sim}{N}\underset{\sim}{K}^{-1}\underset{\sim}{N}')\underset{\sim}{\hat{\tau}} ;$$

(iii) $$\underset{\sim}{B} - \underset{\sim}{N}'\underset{\sim}{R}^{-1}\underset{\sim}{T} = (\underset{\sim}{K} - \underset{\sim}{N}'\underset{\sim}{R}^{-1}\underset{\sim}{N})\underset{\sim}{\hat{\beta}} .$$

It is with the second set of equations that we are primarily concerned; they are called the (reduced) intrablock equations. We shall write them as

$$\underset{\sim}{Q} = \underset{\sim}{C}\hat{\underset{\sim}{\tau}} \ .$$

$\underset{\sim}{Q}$ is the vector of adjusted treatment totals; $\underset{\sim}{C}$ is called the (reduced) intrablock matrix. Some writers use $\underset{\sim}{A}$ rather than $\underset{\sim}{C}$ for the intrablock matrix.

2.3 <u>The analysis of variance table</u>. Under the null hypothesis, $\underset{\sim}{\tau} = \underset{\sim}{0}$, the design is a one-way analysis of variance set-up with blocks as classifications, in which case the sum of squares for blocks is given by

$$S_b = \underset{\sim}{B}'\underset{\sim}{K}^{-1}\underset{\sim}{B} - G^2/n \ .$$

When treatments are included in the model we denote the vector of observations by $\underset{\sim}{Y}$, whence

$$S_e = \underset{\sim}{Y}'\underset{\sim}{Y} - G\hat{\mu} - \underset{\sim}{B}'\hat{\underset{\sim}{\beta}} - \underset{\sim}{T}'\hat{\underset{\sim}{\tau}} \ .$$

The original normal equations give

$$\hat{\underset{\sim}{\beta}} = \underset{\sim}{K}^{-1}\underset{\sim}{B} - \underset{\sim}{K}^{-1}\underset{\sim}{N}'\hat{\underset{\sim}{\tau}} - \underset{\sim}{1}\hat{\mu} \ .$$

Substituting for $\hat{\underset{\sim}{\beta}}$, and recalling that $G = \underset{\sim}{B}'\underset{\sim}{1}$, we have

$$\begin{aligned} S_e &= \underset{\sim}{Y}'\underset{\sim}{Y} - G\hat{\mu} - \underset{\sim}{B}'\underset{\sim}{K}^{-1}\underset{\sim}{B} + \underset{\sim}{B}'\underset{\sim}{K}^{-1}\underset{\sim}{N}'\hat{\underset{\sim}{\tau}} + \underset{\sim}{B}'\underset{\sim}{1}\hat{\mu} - \underset{\sim}{T}'\hat{\underset{\sim}{\tau}} \\ &= (\underset{\sim}{Y}'\underset{\sim}{Y} - G^2/n) - (\underset{\sim}{B}'\underset{\sim}{K}^{-1}\underset{\sim}{B} - G^2/n) - \underset{\sim}{Q}'\hat{\underset{\sim}{\tau}} \ . \end{aligned}$$

The term $\underset{\sim}{Q}'\hat{\underset{\sim}{\tau}}$ represents the decrease in the sum of squares of residuals due to adding treatments to the model, and is usually called the sum of squares for treatments (adjusted for blocks). The hypothesis $\underset{\sim}{\tau} = \underset{\sim}{0}$ is tested by the statistic $F = [\underset{\sim}{Q}'\hat{\underset{\sim}{\tau}} / (t-1)] \div [S_e/(n-t-b+1)]$, which has, under the null hypothesis the $F(t-1, n-t-b+1)$ distribution (see John, 1971).

2.4 <u>The covariance matrix of the adjusted treatment totals</u>. Any solution to the intrablock equations involves the adjusted treatment totals. These totals are clearly not independent, since $\mathbf{1}'\mathbf{Q} = 0$. It will be necessary later to use the covariance matrix $\text{cov}(\mathbf{Q})$, which will now be shown to be $\mathbf{C}\sigma^2$. Consider the vector $\mathbf{Z} = (\mathbf{T}', \mathbf{B}')'$ with $t + b$ random variables:

$$\text{cov}(\mathbf{Z}) = \begin{bmatrix} \mathbf{R} & \mathbf{N} \\ \mathbf{N}' & \mathbf{K} \end{bmatrix} \sigma^2 , \qquad \mathbf{Q} = (\mathbf{I} , -\mathbf{N}\mathbf{K}^{-1})\mathbf{Z} .$$

Then

$$\text{cov}(\mathbf{Q}) = [\mathbf{I} , -\mathbf{N}\mathbf{K}^{-1}] \begin{bmatrix} \mathbf{R} & \mathbf{N} \\ \mathbf{N}' & \mathbf{K} \end{bmatrix} \begin{bmatrix} \mathbf{I} \\ -\mathbf{K}^{-1}\mathbf{N}' \end{bmatrix} \sigma^2 = \mathbf{C}\sigma^2 .$$

2.5 Solving the normal equations. It would be convenient if we could write $\hat{\boldsymbol{\tau}} = \mathbf{C}^{-1}\mathbf{Q}$, but $\mathbf{C}\mathbf{1} = \mathbf{R}\mathbf{1} - \mathbf{N}\mathbf{K}^{-1}\mathbf{N}'\mathbf{1} = \mathbf{R}\mathbf{1} - \mathbf{N}\mathbf{1} = \mathbf{0}$, so that $\mathbf{C}$ is singular; fortunately, since $\mathbf{Q}'\mathbf{1} = 0$, the intrablock equations are consistent. We shall confine our investigations to designs for which the rank of $\mathbf{C}$ is $t - 1$. These are called connected designs and for them, as we shall see in the next paragraph, all contrasts in the treatments, i.e., all linear combinations $\mathbf{c}'\boldsymbol{\tau}$, where $\mathbf{c}'\mathbf{1} = 0$, have unique least square estimates.

Let $\mathbf{G}$ and $\mathbf{H}$ be any two generalized inverses of $\mathbf{C}$, by which we mean that they are square matrices of order $t$ such that $\mathbf{GQ}$ and $\mathbf{HQ}$ are both solution vectors to the intrablock equations. Then $\mathbf{Q} = \mathbf{C}\hat{\boldsymbol{\tau}}$ implies $\mathbf{Q} = \mathbf{CGQ}$ and $\mathbf{Q} = \mathbf{CHQ}$ for all vectors $\mathbf{Q}$ of adjusted treatment totals so that $\mathbf{C}(\mathbf{G} - \mathbf{H})\mathbf{Q} = \mathbf{0}$. It follows that $(\mathbf{G} - \mathbf{H})\mathbf{Q}$ can be written $a\mathbf{1}$, where $a$ is a scalar, which may be zero. Let $\mathbf{c}$ be a vector such that $\mathbf{c}'\mathbf{1} = 0$. The two estimates of $\mathbf{c}'\boldsymbol{\tau}$ are $\mathbf{c}'\mathbf{GQ}$ and $\mathbf{c}'\mathbf{HQ}$; but $\mathbf{c}'\mathbf{GQ} - \mathbf{c}'\mathbf{HQ} = \mathbf{c}'(\mathbf{G} - \mathbf{H})\mathbf{Q} = a\mathbf{c}'\mathbf{1} = 0$, so that the estimate of $\mathbf{c}'\boldsymbol{\tau}$ is unique.

We shall be considering symmetric generalized inverses of the form $\boldsymbol{\Omega} = (\mathbf{C} + a\mathbf{J})^{-1}$ where $a$ is any convenient nonzero scalar. We have seen that $\mathbf{1}$ is a latent vector of $\mathbf{C}$ corresponding to the simple latent root zero; $\mathbf{1}$ is also a latent vector of $\mathbf{C} + a\mathbf{J}$ and $at$ is the latent root. If $\psi$ is any other latent root of $\mathbf{C}$ and $\mathbf{x}$ is a corresponding latent vector, $\mathbf{x}'\mathbf{1} = 0$ and so $\psi$ is also a latent root of $\mathbf{C} + a\mathbf{J}$ with the vector $\mathbf{x}$. It follows that all the latent roots of $\mathbf{C} + a\mathbf{J}$ are nonzero. Furthermore the latent roots of $\boldsymbol{\Omega}$ are the simple root $(at)^{-1}$ with $\mathbf{1}$

as vector, and the reciprocals of the other nonzero latent roots of $\underset{\sim}{C}$ with the same multiplicities and the same latent vectors.

For any such generalized inverse we have

$$(\underset{\sim}{C} + a\underset{\sim}{J})\underset{\sim}{\Omega} = \underset{\sim}{I}$$

so that

$$\underset{\sim}{C}\underset{\sim}{\Omega} = \underset{\sim}{I} - a\underset{\sim}{J}\underset{\sim}{\Omega} = \underset{\sim}{I} - \underset{\sim}{J}/t \quad .$$

Recalling that $\operatorname{cov}\underset{\sim}{Q} = \underset{\sim}{C}\sigma^2$ and that, since $\underset{\sim}{c}'\underset{\sim}{1} = 0$, $\underset{\sim}{J}\underset{\sim}{c} = \underset{\sim}{0}$, we obtain

$$V(\underset{\sim}{c}'\hat{\underset{\sim}{\tau}}) = V(\underset{\sim}{c}'\underset{\sim}{\Omega}\underset{\sim}{Q}) = \underset{\sim}{c}'\underset{\sim}{\Omega}\underset{\sim}{C}\underset{\sim}{\Omega}\underset{\sim}{c}\sigma^2 = \underset{\sim}{c}'\underset{\sim}{\Omega}\underset{\sim}{c}\sigma^2 \quad .$$

We remark, also, that $\underset{\sim}{\Omega}$ has the property that $\underset{\sim}{C}\underset{\sim}{\Omega}\underset{\sim}{C} = \underset{\sim}{C}$.

The solution vector, $\hat{\underset{\sim}{\tau}}$, obtained with $\underset{\sim}{\Omega}$ has the constraint, or side condition, $\underset{\sim}{1}'\hat{\underset{\sim}{\tau}} = 0$, since

$$\underset{\sim}{1}'\hat{\underset{\sim}{\tau}} = \underset{\sim}{1}'\underset{\sim}{\Omega}\underset{\sim}{Q} = (at)^{-1}\underset{\sim}{1}'\underset{\sim}{Q} = 0 \quad .$$

Another generalized inverse, which differs from $\underset{\sim}{\Omega}$ when the treatments are not all replicated the same number of times, was introduced by Tocher (1952). We denote it by $\underset{\sim}{\Omega}^*$;

$$\underset{\sim}{\Omega}^{*-1} = \underset{\sim}{C} + \underset{\sim}{r}\underset{\sim}{r}'/n \quad .$$

For this generalized inverse

$$\underset{\sim}{\Omega}^{*-1}\underset{\sim}{1} = \underset{\sim}{C}\underset{\sim}{1} + n^{-1}\underset{\sim}{r}\underset{\sim}{r}'\underset{\sim}{1} = \underset{\sim}{r}$$

and

$$\underset{\sim}{\Omega}^*\underset{\sim}{r} = 1 \;;$$

$$\underset{\sim}{C}\underset{\sim}{\Omega}^* = \underset{\sim}{I} - n^{-1}\underset{\sim}{r}\underset{\sim}{1}' \;, \quad \underset{\sim}{C}\underset{\sim}{\Omega}^*\underset{\sim}{C} = \underset{\sim}{C} \quad .$$

The solution vector obtained with $\underset{\sim}{\Omega}^*$ is subject to the side condition $\underset{\sim}{r}'\hat{\underset{\sim}{\tau}} = 0$, since

$$\underset{\sim}{r}'\hat{\underset{\sim}{\tau}} = \underset{\sim}{r}'\underset{\sim}{\Omega}^*\underset{\sim}{Q} = \underset{\sim}{1}'\underset{\sim}{Q} = 0 \quad .$$

An attractive feature of the use of $\underset{\sim}{\Omega}^*$ rather than $\underset{\sim}{\Omega}$ when the $r_i$ are not equal is that, when coupled with a similar side condition, $\underset{\sim}{1}'\underset{\sim}{K}\underset{\sim}{\beta} = 0$, on the block effects, it leads to the estimate $\hat{\mu} = G/n$.

The following numerical example, which has been constructed with zero errors, illustrates some of the properties of $\underset{\sim}{\Omega}$ and $\underset{\sim}{\Omega}^*$ and the differences between some of the estimates.

2.6 A numerical example. The data are given below. There are five blocks; $k_1 = k_3 = 4$; $k_2 = k_4 = k_5 = 2$; $\underset{\sim}{r}' = (6, 4, 4)$. The varieties and the 'observations' are:

| | | | | |
|---|---|---|---|---|
| A 12, | A 12, | B 14, | C 16; | $B_1 = 54$; |
| B 16, | C 18; | | | $B_2 = 34$; |
| A 16, | A 16, | B 18, | C 20; | $B_3 = 70$; |
| A 18, | B 20; | | | $B_4 = 38$; |
| A 20, | C 24; | | | $B_5 = 44$ . |

$T_1 = 94$, $T_2 = 68$, $T_3 = 78$, $G = 240$, $\underset{\sim}{Q}' = (-9, 1, 8)$.

We now obtain

$$\underset{\sim}{C} = \frac{1}{2}\begin{bmatrix} 6 & -3 & -3 \\ -3 & 5 & -2 \\ -3 & -2 & 5 \end{bmatrix} .$$

Using the first solution to the normal equations

$$\underset{\sim}{\Omega}^{-1} = \underset{\sim}{C} + 3\underset{\sim}{J}/2 = \frac{1}{2}\begin{bmatrix} 9 & 0 & 0 \\ 0 & 8 & 1 \\ 0 & 1 & 8 \end{bmatrix}, \quad \underset{\sim}{\Omega} = \frac{2}{63}\begin{bmatrix} 7 & 0 & 0 \\ 0 & 8 & -1 \\ 0 & -1 & 8 \end{bmatrix},$$

so that $\hat{\underset{\sim}{\tau}}' = (-2, 0, 2)$. If we impose the side condition $\underset{\sim}{1}'\hat{\underset{\sim}{\beta}} = 0$ in the equations $\underset{\sim}{B} = \underset{\sim}{K}\underset{\sim}{1}\mu + \underset{\sim}{K}\hat{\underset{\sim}{\beta}} + \underset{\sim}{N}'\hat{\underset{\sim}{\tau}}$, we have

$$\underset{\sim}{1}'\underset{\sim}{K}^{-1}\underset{\sim}{B} = \underset{\sim}{1}'\underset{\sim}{1}\hat{\mu} + \underset{\sim}{1}'\underset{\sim}{K}^{-1}\underset{\sim}{N}'\hat{\underset{\sim}{\tau}} ,$$

whence $\hat{\mu} = 18$, $\hat{\underset{\sim}{\beta}}' = (-4, -2, 0, 2, 4)$. On the other hand

$$\underset{\sim}{\Omega^{*}}{}^{-1} = \frac{1}{14} \begin{bmatrix} 78 & 3 & 3 \\ 3 & 51 & 2 \\ 3 & 2 & 51 \end{bmatrix} , \quad \underset{\sim}{\Omega^{*}} = \frac{1}{294} \begin{bmatrix} 53 & -3 & -3 \\ -3 & 81 & -3 \\ -3 & -3 & 81 \end{bmatrix}$$

with estimates $\underset{\sim}{\tilde{\tau}}' = (-12/7, 2/7, 16/7)$. Imposing the side condition $1'K\underset{\sim}{\tilde{\beta}}$ = 0 leads to $\tilde{\mu} = 120/7$ and $\underset{\sim}{\tilde{\beta}}' = (-24, -10, 4, 18, 32)/7$. We note that, although $\underset{\sim}{\hat{\tau}} \neq \underset{\sim}{\tilde{\tau}}$ , nevertheless $\hat{\tau}_2 - \hat{\tau}_1 = 2 = \tilde{\tau}_2 - \tilde{\tau}_1$ and $\hat{\tau}_3 - \hat{\tau}_1 = \tilde{\tau}_3 - \tilde{\tau}_1$. Indeed $\underset{\sim}{\tilde{\tau}} = \underset{\sim}{\hat{\tau}} + (2/7)1$ and for any vector $\underset{\sim}{c}$ such that $\underset{\sim}{c}'1 = 0$, we have $\underset{\sim}{c}'\underset{\sim}{\hat{\tau}} = \underset{\sim}{c}'\underset{\sim}{\tilde{\tau}}$. However, the estimates of $\mu + \tau_i$ are not unique. We have

$$(\hat{\mu}1 + \underset{\sim}{\hat{\tau}})' = (16, 18, 20), \quad (\tilde{\mu}1 + \underset{\sim}{\tilde{\tau}})' = (108, 122, 136)/7 .$$

2.7  Proper, binary, equireplicate designs. If all the blocks have the same number of plots, i.e., $k_j = k$ for all $j$, a design is called proper. If $r_i = r$ for each $i$, it is equireplicate. If $n_{ij}$ takes only the values zero or one, it is binary. The subsequent chapters will be concerned almost entirely with proper binary equireplicate designs, and we shall henceforth assume that to be the case without saying so.

For proper binary equireplicate designs,

$$\underset{\sim}{C} = r\underset{\sim}{I} - \underset{\sim}{NN}'/k \; ;$$

$\underset{\sim}{NN}'$ is called the concordance matrix. The diagonal elements of $\underset{\sim}{NN}'$ are each equal to $r$. The off-diagonal element in the $h^{th}$ row and the $i^{th}$ column is equal to the number of blocks in which the $h^{th}$ and $i^{th}$ varieties both appear; it is sometimes written as $\lambda_{hi}$.

Each row of $\underset{\sim}{NN}'$ sums to $r + \underset{h \neq i}{\Sigma} \lambda_{hi} = r + r(k - 1) = rk$, so that $\underset{\sim}{NN}'1 = rk1$. For a connected design $rk$ is a simple root of $\underset{\sim}{NN}'$, corresponding to the latent vector $\underset{\sim}{1}$ and to the zero root of $\underset{\sim}{C}$. Indeed if $\theta$ is a latent root of $\underset{\sim}{NN}'$ with latent vector $\underset{\sim}{x}$, then $\psi = r - \theta/k$ is a root of $\underset{\sim}{C}$ with $\underset{\sim}{x}$ as a vector; it follows that $\theta$ and $\psi$ have the same multiplicity.

This connection between the latent roots and vectors of $\underset{\sim}{\Omega}$, $\underset{\sim}{C}$, and $\underset{\sim}{NN}'$ is important. We shall be interested in designs which have a certain structure for $\underset{\sim}{\Omega}$, since $V(\underset{\sim}{c}'\underset{\sim}{\hat{\tau}}) = \underset{\sim}{c}'\underset{\sim}{\Omega}\underset{\sim}{c}$. This structure will be reflected in the structure of $\underset{\sim}{NN}'$, so that we shall be able to work in terms of the concordance matrix rather than $\underset{\sim}{C}$ or $\underset{\sim}{\Omega}$ .

# Chapter Three

# BALANCED INCOMPLETE BLOCK DESIGNS

3.1 Introduction. As we indicated at the end of Chapter 2, we shall now confine ourselves to proper binary equireplicate designs. If $k < t$ we cannot have a situation in which each pair of treatments appears together in the same block $r$ times. We must therefore seek a compromise. Can we find designs in which each pair of treatments appears together in the same number, $\lambda(<r)$, of blocks? The answer is yes. These are the balanced incomplete block designs (BIBD) which were introduced by Yates in 1936. The design for four treatments which was the example in the first chapter is a BIBD.

A balanced incomplete block design has five parameters $t, b, r, k, \lambda$. For the mathematician we can define it as a collection of $b$ subsets of size $k$ from a set of $t$ elements, such that (i) each element appears in exactly $r$ subsets, and (ii) each pair of elements appears in exactly $\lambda$ subsets.

3.2 Relations between the parameters. The five parameters are not independent. They are related by two equalities:

(i) $rt = bk = n$ (the total number of plots),

(ii) $\lambda(t-1) = r(k-1)$.

To establish the latter equality we note that any treatment appears in $r$ blocks in which there is a total of $r(k-1)$ other plots. These plots must contain each of the other $(t-1)$ treatments $\lambda$ times each.

There is also an inequality called Fisher's inequality (Fisher, 1940):

(iii) $b \geq t$.

The following proof was given by Bose (1949).

We note that in the concordance matrix $\underset{\sim}{N}\underset{\sim}{N}'$ each of the diagonal elements is $r$, and each of the other elements is $\lambda$; $\underset{\sim}{N}\underset{\sim}{N}' = (r - \lambda)\underset{\sim}{I} + \lambda\underset{\sim}{J}$. We need the following lemma.

Lemma 3.2.1. Let $\underset{\sim}{P}$ be a square matrix $(p_{ij})$ with $n$ rows and columns having $p_{ii} = q$ for all $i$ and $p_{ij} = s$ for $i \neq j$, i.e., $\underset{\sim}{P} = (q-s)\underset{\sim}{I}_n + s\underset{\sim}{J}_n$. Then $\underset{\sim}{P} = (q-s)^{n-1}(q + (n-1)s)$.

Proof: We transform $\underset{\sim}{P}$ to another matrix $\underset{\sim}{U}$ in two stages. First subtract row 2 from row 1, then subtract row 3 from row 2, etc.; second, add column 1 to column 2, then add the new column 2 to column 3, etc. We now evaluate $|\underset{\sim}{P}| = |\underset{\sim}{U}|$ as the product of the diagonal elements of $\underset{\sim}{U}$.

Applying this lemma the determinant $|\underset{\sim}{N}\underset{\sim}{N}'|$ is now evaluated as

$$|\underset{\sim}{N}\underset{\sim}{N}'| = (r-\lambda)^{t-1}(r + (t-1)\lambda) = (r-\lambda)^{t-1}rk,$$

which is positive since $\lambda < r$. Thus the rank of $\underset{\sim}{N}\underset{\sim}{N}'$ is $t$ and so is the rank of $\underset{\sim}{N}$. But $r(\underset{\sim}{N}) = \min(t,b)$, since $\underset{\sim}{N}$ has $t$ rows and $b$ columns, so that $t = \min(t,b)$, and $b \geq t$.

3.3 Intrablock analysis of balanced incomplete block designs. The intrablock matrix is

$$\underset{\sim}{C} = r\underset{\sim}{I} - \underset{\sim}{N}\underset{\sim}{N}'/k = r\underset{\sim}{I} - ((r-\lambda)\underset{\sim}{I} + \lambda\underset{\sim}{J})/k$$
$$= \lambda t\underset{\sim}{I}/k - \lambda\underset{\sim}{J}/k.$$

It is obviously convenient to take $a = \lambda/k$ so that

$$\underset{\sim}{\Omega}^{-1} = \underset{\sim}{C} + a\underset{\sim}{J} = \lambda t\underset{\sim}{I}/k \quad \text{and} \quad \underset{\sim}{\Omega} = k\underset{\sim}{I}/(\lambda t).$$

Then $\hat{\tau}_h = kQ_h/(\lambda t)$ for all $h$, and $V(\hat{\tau}_h - \hat{\tau}_i) = 2k\sigma^2/(\lambda t)$ for all pairs. The latter property can be used for a more general definition of balanced designs including those which are not proper binary equireplicate. It should be kept in mind, however, that the expression "balanced incomplete block design" is only used for the designs that we have just described.

Definition. An incomplete block design is said to be variance-balanced if the variance of a simple comparison, $V(\hat{\tau}_h - \hat{\tau}_i)$, is the same for all pairs of treatments. John (1964, 1976) and Hedayat and John (1974) have considered variance-balanced designs with unequal block sizes or unequal numbers of replicates.

3.4 Some series of balanced incomplete block designs. The simplest examples of balanced incomplete block designs are the sets of all k-tuples from the $t$ treatments. These are called the unreduced designs. For them

$$b = \binom{t}{k}, \quad r = \binom{t-1}{k-1}, \quad \lambda = \binom{t-2}{k-2} .$$

Yates introduced two series of designs based upon sets of mutually orthogonal latin squares. A latin square is a square array of side $s$ in which $s$ symbols (traditionally latin letters) are written $s$ times each in such a way that each letter appears exactly once in each row and exactly once in each column. Two latin squares of side $s$ are said to be orthogonal if, when we superimpose one square on the other, each symbol of the first square coincides exactly once with each symbol of the second square; such a pair is then said to form a graeco-latin square (as if we had used greek letters for one square and latin letters for the other). It is well known that there does not exist a graeco-latin square of side six. A set of squares is said to be a set of mutually orthogonal latin squares (MOLS) if they are pairwise orthogonal. We shall show later that if $s$ is a prime or a power of a prime we can always find a set of $(s-1)$ MOLS.

We illustrate Yates' orthogonal series by using the set of three mutually orthogonal squares of side four.

| | | | | | | | | | | | | | |
|---|---|---|---|---|---|---|---|---|---|---|---|---|---|
| A | B | C | D | | A | B | C | D | | A | B | C | D |
| B | A | D | C | | C | D | A | B | | D | C | B | A |
| C | D | A | B | | D | C | B | A | | B | A | D | C |
| D | C | B | A | | B | A | D | C | | C | D | A | B |

To obtain the OS1 design for $t = 16$ we first write down a square array of sixteen numbers and then transpose it. Letting the numbers correspond to the treatments and the rows of the arrays to blocks, we now have the first eight blocks of a design with $k = 4$.

| | | | | | | | | | |
|---|---|---|---|---|---|---|---|---|---|
| I | 1 | 2 | 3 | 4 | V | 1 | 5 | 9 | 13 |
| II | 5 | 6 | 7 | 8 | VI | 2 | 6 | 10 | 14 |
| III | 9 | 10 | 11 | 12 | VII | 3 | 7 | 11 | 15 |
| IV | 13 | 14 | 15 | 16 | VIII | 4 | 8 | 12 | 16 |

To obtain the next four blocks we superimpose the array upon the first latin square. The ninth block contains the four treatments that appear with the letter A, the tenth block contains those treatments that go with B, and so on. Blocks 13 through 16 are obtained by superimposing the array on the second square; blocks 17 through 20 come from the third square.

| | | | | | | | | | | | | | | |
|---|---|---|---|---|---|---|---|---|---|---|---|---|---|---|
| IX | 1 | 6 | 11 | 16 | XIII | 1 | 7 | 12 | 14 | XVII | 1 | 8 | 10 | 15 |
| X | 2 | 5 | 12 | 15 | XIV | 2 | 8 | 11 | 13 | XVIII | 2 | 7 | 9 | 16 |
| XI | 3 | 8 | 9 | 14 | XV | 3 | 5 | 10 | 16 | XIX | 3 | 6 | 12 | 13 |
| XII | 4 | 7 | 10 | 13 | XVI | 4 | 6 | 9 | 15 | XX | 4 | 5 | 11 | 14 |

This method gives designs for $t = s^2$, $b = s(s+1)$, $r = s+1$, $k = s$, $\lambda = 1$, whenever we can find a set of $(s-1)$ MOLS.

The OS1 design can be divided into $s+1$ sets of $s$ blocks each in such a way that each set consists of a single replicate of all $s^2$ treatments. Such a design is said to be resolvable.

The OS2 design is obtained from the OS1 design in the following way. To each block in the first set, we add a new treatment 17. To each block in the second set, we add 18, and so on. Finally we add a new block consisting of the five new treatments 17, 18, 19, 20, 21. This procedure gives a series of designs with $t = b = s^2 + s+1$, $r = k = s+1$, $\lambda = 1$, whenever the corresponding OS1 design exists.

Definition. A binary proper equireplicate incomplete block design for which $t = b$, $r = k$ is said to be a symmetric design.

3.5 The derivation of sets of mutually orthogonal squares. We present now the method given by Bose (1938) for obtaining sets of $(s-1)$ MOLS when $s$ is a prime or a power of a prime. We write the elements of the Galois field $GF(s)$ as $g_0 = 0$, $g_1 = 1$, $g_2 = x$, $g_3 = x^2, \ldots, g_{s-1} = x^{s-2}$, where $x$ is a primitive element of the field, and obtain a set of $(s-1)$ MOLS in the following way.

We number the rows and columns of the squares $0, 1, 2, \ldots, s-1$. The first row of each square is $g_0, g_1, \ldots, g_{s-1}$. The element in the $h^{th}$ row and the $i^{th}$ column of the $j^{th}$ square is $g_j g_h + g_i$. This says that in the first square we obtain the $h^{th}$ row by adding $g_h$ to each element of the first row; we obtain the subsequent squares by rotating the rows other than the first cyclically.

The elements of $GF(2^2)$ are $0, 1, x, 1+x$, with arithmetic mod 2 and the restriction $x^2 = 1 + x$. In the example of Section 3.3, we wrote A for 0, B for 1, C for x, D for 1 + x.

We must establish two properties to validate our method.

(i) Each square is a latin square.

Suppose that in the $h^{th}$ row of the $j^{th}$ square the same element appears in both the $i^{th}$ and the $m^{th}$ columns. Then $g_j g_h + g_i = g_j g_h + g_m$. This implies that $g_i = g_m$, which is false.

(ii) The squares are mutually orthogonal.

Suppose that in the $j^{th}$ and $n^{th}$ squares the same elements appear in both the $h^{th}$ row and the $i^{th}$ column and the $k^{th}$ row and the $m^{th}$ column. This implies

$$g_h g_j + g_i = g_k g_j + g_m ,$$

$$g_h g_n + g_i = g_k g_n + g_m ;$$

subtracting, we have $(g_h - g_k)(g_j - g_n) = 0$, which is only possible if $g_h = g_k$ or $g_j = g_n$.

3.6 <u>The nonexistence of a design</u>. We established in Section 3.2 some necessary conditions for the parameters of a BIBD. We may ask if they are sufficient. The following counterexample shows that they are not.

We can retrace our steps and show that, starting with a BIBD with $t = s^2$, $b = s(s+1)$, $r = s + 1$, $k = s$ and $\lambda = 1$, we can construct a set of mutually orthogonal squares of size s. However, there does not exist even a pair of squares of side six. Hence, the design with $t = 36$, $b = 42$, $r = 7$, $k = 6$, $\lambda = 1$ does not exist.

Bruck and Ryser (1949) have proved the following nonexistence theorem. Let $s \equiv 1$ or $2 \bmod 4$. Let q be the square free part of s, by which we mean the quotient after any factors which are squares have been divided out. If there is a prime p, such that $p \equiv 3 \bmod 4$, which divides q, then a full set of $s - 1$ MOLS of side s does not exist.

For example, since 14 is divisible by 7, there is not a set of thirteen MOLS of side $s = 14$. It follows that the orthogonal series designs for $t = 196$, $b = 210$, $r = 15$, $k = 14$, $\lambda = 1$, and for $t = b = 211$, $r = k = 15$, $\lambda = 1$ do not exist.

Other nonexistence theorems will be given in Chapter 4.

# Chapter Four

# THE CONSTRUCTION OF BALANCED INCOMPLETE BLOCK DESIGNS

4.1 Introduction. Given a set of parameters $t, b, r, k, \lambda$, all integers, which satisfy the three necessary conditions (i) $bk = rt$, (ii) $\lambda(t-1) = r(k-1)$, (iii) $b \geq t$, does there exist a corresponding design? We have given in Chapter 3 some recipes and some examples of nonexistence. In every case we must either produce a design, or, at least, a recipe for its construction, or else prove nonexistence. For some sets of parameters we are still not able to answer the question, and these cases remain as unsolved problems for the researcher.

Bose (1939) considered all sets of parameters satisfying the three necessary conditions above with $r \leq 10$ and $k \leq 10$. He left twelve cases unsolved. Later workers solved ten of these cases, but two remain open. We do not know whether there exist designs with

(i) $t = 46$, $b = 69$, $r = 9$, $k = 6$, $\lambda = 1$,

(ii) $t = 51$, $b = 85$, $r = 10$, $k = 6$, $\lambda = 1$.

On the other hand Hanani (1961,1965) has shown that the above conditions are also sufficient when $k = 3$, or 4, and when $k = 5$ and $\lambda = 1$, 4 or 20.

If D is a binary design with $t$ treatments in $b$ blocks of size $k$, the complementary design D' is a design for $t$ treatments in $b$ blocks of size $t - k$. The $j^{th}$ block of D' consists of those treatments that do not appear in the $j^{th}$ block of D.

Theorem 4.1.1. If D is a BIBD so is D'.

Proof: Consider two treatments $x$ and $y$. They appear together in D' in exactly those blocks which are the complements of the blocks in D in which neither x nor y appears. Denote the number of such blocks by $n(x,y)$. The design D has $\lambda$ blocks containing both x and y, $r - \lambda$ blocks containing x but not y, and $r - \lambda$ blocks which contain y but not x. It follows that $n(x,y) = b - \lambda - 2(r - \lambda) = b - 2r + \lambda$, which is the same for all choices of the pair x, y, so that D' is a BIBD.

The importance of this theorem is that in the search for balanced incomplete block designs we need only consider sets of parameters with $2k \leq t$.

A listing of BIB designs with $t \leq 100$, $b \leq 100$, $r \leq 15$, $k \leq 15$, and one solution for each case in which a solution is known to exist is found in the book by Raghavarao (1971 pp. 91 et seq., table 5.10.1). In the sections which follow we shall discuss various aspects of Bose's method of constructing BIB designs by the cyclic development of initial blocks. In the first part of his 1938 paper Bose derived some designs from finite geometries, and these are mentioned by Raghavarao in his listing. Rao (1946) showed that all the designs obtained from finite geometries can also be obtained by developing initial blocks, and so there is no need to make them the topic of separate discussion.

4.2 Difference sets. We begin with an example of a symmetric design with $t = b = 7$, $r = k = 3$, $\lambda = 1$. We denote the treatments by 0, 1, 2, 3, 4, 5, 6 which are the elements of the finite field GF(7) with arithmetic mod 7. Consider the set 0, 1, 3. The differences between pairs of elements reduced mod 7 are $1 - 0 = 1$, $3 - 0 = 3$, $3 - 1 = 2$, $0 - 1 = 6$, $0 - 3 = 4$, $1 - 3 = 5$; this set of six differences contains each nonzero element of the set of treatments exactly once.

Definition: Let $0, 1, \ldots, t - 1$ be the elements of an Abelian group under addition. Let A be a subset of k elements such that the $k(k - 1)$ differences between members of A comprise all the nonzero elements of the group exactly $\lambda$ times each. Then A is said to be a difference set or, more formally, a perfect difference set.

We take the difference set 0, 1, 3 as the initial block of our design and obtain the other six blocks by developing the initial block cyclically. The $j^{th}$ block of the design $(j = 0, 1, 2, 3, 4, 5, 6)$ is obtained by adding j to each element of the initial block (and, in this case, reducing mod 7). The seven blocks of the design are thus:

0, 1, 3; 1, 2, 4; 2, 3, 5; 3, 4, 6; 4, 5, 0; 5, 6, 1; 6, 0, 2.

Consider the $i^{th}$ treatment. In the block in which it stands in the first position it appears with $i + 1$ and $i + 3$; in the block in which it holds the second position it appears with $i + 6$ and $i + 2$; it shares its other appearance with the remaining two treatments, $i + 4$ and $i + 5$. Thus we have here a design in which each treatment appears with each other treatment exactly once. The reader may be interested in showing that this design is isomorphic to the OS2 design for $s = 2$, which was obtained in Chapter 3.

The following theorem is now obvious and its proof will not be given.

Theorem 4.2.1. Let A be a difference set of size k for a set of t elements such that each nonzero difference occurs $\lambda$ times. If the set A is taken as an initial block and developed cyclically it generates a symmetric BIBD with parameters t, k, $\lambda$.

In most cases the Abelian group will either be the set of integers $0, 1, \ldots, t - 1$ with arithmetic mod t or the Galois field $GF(p^m)$, but this is not necessary. The following example illustrates this point: the parameters of the design are $t = 15$, $k = 7$, $\lambda = 3$.

Each treatment is represented by a pair of integers $(x,y)$, $x = 0, 1, 2, 3, 4$ and $y = 0, 1, 2$. Addition is defined by $(x_1, y_1) + (x_2, y_2) = (x_1 + x_2, y_1 + y_2)$, where $x_1 + x_2$ is reduced mod 5 and $y_1 + y_2$ is reduced mod 3. Then 00, 10, 40, 01, 21, 31, 02 is a difference set with $\lambda = 3$. The Abelian group used here is the direct product of the two groups 0, 1, 2, 3, 4 mod 5 and 0, 1, 2 mod 3. This example comes from Bose (1939).

A well known result from number theory provides us with a useful series of difference sets. Let t be a prime of the form $4m + 3$. We consider the integers 0, 1, 2,... mod t; $2m + 1$ of the nonzero integers are squares (mod t); the remaining $2m + 1$ positive integers are not squares. The two sets are called the quadratic residues and the nonquadratic residues, respectively. The theorem states that each of these two sets is a difference set. They give us a family of symmetric BIB designs with $t = 4m + 3$, $k = 2m + 1$ and $\lambda = m$; the complementary designs are also BIB designs with $k = 2m + 2$ and $\lambda = m + 1$.

In the case $t = 7$ we may take 3 as a primitive element of the field; the quadratic residues are $3^2 = 2$, $3^4 = 4$ and $3^6 = 1$. Thus 1, 2, 4 is a difference set; subtracting 1 from each element gives the set 0, 1, 3 that we used at the beginning of this section.

An obvious extension of this method of obtaining a symmetric design from a single initial block is to use $u$ initial blocks and obtain a balanced design with $b = tu$ blocks of size $k$. It will be necessary and sufficient that among the $uk(k-1)$ differences in the $u$ initial blocks each nonzero value shall occur exactly $\lambda$ times. We can, for example, obtain a design for $t = 11$, $b = 55$, $r = 15$, $k = 3$, $\lambda = 3$ by developing the following five initial blocks:

0, 1, 10; 0, 2, 9; 0, 3, 8; 0, 4, 7; 0, 5, 6.

4.3 The point at infinity. Returning to our example with $t = 7$, $k = 3$, $\lambda = 1$, the complement of the initial block is 2, 4, 5, 6. It is also a difference set with $k = 4$ and $\lambda = 2$. We now add to the initial block 0, 1, 3 an eighth treatment denoted by $\infty$, a symbol which represents infinity in the sense that it remains unchanged as the cyclical development proceeds. To a geometer - we mentioned earlier that some of the designs by Bose (1939) were derived from finite geometries - it corresponds to the point at infinity. Developing both blocks gives a design for $t = 8$, $b = 14$, $r = 7$, $k = 4$, $\lambda = 3$:

| | | | | | | | | |
|---|---|---|---|---|---|---|---|---|
| 0 | 1 | 3 | ∞ | | 2 | 4 | 5 | 6 |
| 1 | 2 | 4 | ∞ | | 3 | 5 | 6 | 0 |
| 2 | 3 | 5 | ∞ | | 4 | 6 | 0 | 1 |
| 3 | 4 | 6 | ∞ | | 5 | 0 | 1 | 2 |
| 4 | 5 | 0 | ∞ | | 6 | 1 | 2 | 3 |
| 5 | 6 | 1 | ∞ | | 0 | 2 | 3 | 4 |
| 6 | 0 | 2 | ∞ | | 1 | 3 | 4 | 5 |

This method of obtaining designs is formalized in the following theorem, which is stated without proof.

Theorem 4.3.1. Let $G$ be an Abelian group under addition with $t - 1$ elements and let these elements together with the addition of $\infty$ denote the treatments. Let $A$ denote a set of $s + u$ initial blocks, each of size $k$, subject to the following conditions:

(i) $s$ of the blocks each contain $k$ distinct elements of $G$;

(ii) each of the remaining $u$ blocks contains $\infty$ together with $k - 1$ elements of $G$;

(iii) ignoring $\infty$, the collection of differences between the elements of $A$ in the $s+u$ initial blocks contains each nonzero member of $G$ the same number, $\lambda$, of times;

(iv) $\lambda = u(k-1)$.

Then, developing the blocks of $A$ cyclically gives a BIB design.

Corollary 4.3.2. Let $t-1 = 4m+3$ be a prime or a power of a prime. There exists a BIB design for $r$ treatments with $b = 8m+6$, $4m+3$, $k = 2m+2$, $\lambda = 2m+1$.

Proof. Let $G$ be the Galois field with $t-1$ elements, and let $x$ be a primitive element of the field. We take as the two initial blocks $(1, x^2, x^4, \ldots, x^{4m}, \infty)$, $(0, x, x^3, \ldots, x^{4m+1})$.

Example: let $t = 12$. The initial blocks are $(1, 4, 5, 9, 3, \infty)$ and $(0, 2, 8, 10, 7, 6)$.

4.4 The fundamental theorems of Bose. Theorems 4.2.1 and 4.3.1 were generalized by Bose (1939) in his two fundamental theorems of differences. He considers $t = mn$ treatments which are divided into $m$ groups of size $n$. Each group is then associated with an element of an Abelian group. The idea of a difference set is generalized to the concept of symmetrically repeated differences.

The reader who wishes to follow this topic further is referred to Bose's original paper or to the later books by John (1971) and Raghavarao (1971).

4.5 A nonexistence theorem for symmetric designs. The following theorem was discovered independently and at about the same time by Schutzenberger (1949), Shrikhande (1950), and Chowla and Ryser (1950). It is interesting to students to note that it was unknown to Bose in 1939, and it provided, belatedly, negative solutions to two of the cases that Bose had left unsolved.

Theorem 4.5.1. If in a symmetric BIBD $t$ is even, then $k - \lambda$ must be a square.

Proof: We have seen in the previous chapter that for any BIBD the determinant $|\underset{\sim}{N}\underset{\sim}{N}'| = (r - \lambda)^{t-1} rk$. For a symmetric design this becomes $|\underset{\sim}{N}|^2 = (k - \lambda)^{t-1} k^2$. The right side must be a square. Hence, if $t$ is even, $k - \lambda$ must be a square.

The two symmetric designs in Bose's list which were shown by this theorem to be impossible are $t = 22$, $k = 7$, $\lambda = 2$ and $t = 46$, $k = 10$, $\lambda = 2$.

## 4.6 Block section and block intersection

Theorem 4.6.1. Any two blocks of a symmetric BIBD have exactly $\lambda$ treatments in common.

Proof: For any symmetric equireplicate design (by which we mean a design with $b = t$, and $r = k$, whether balanced or not)

$$\underset{\sim}{N}\,\underset{\sim}{J}_t = k\,\underset{\sim}{J}_t = \underset{\sim}{J}_t\,\underset{\sim}{N}.$$

For a symmetric BIBD $\underset{\sim}{N}\underset{\sim}{N}' = (k - \lambda)\underset{\sim}{I} + \lambda\underset{\sim}{J}$.
Then $\underset{\sim}{N}\underset{\sim}{N}'\underset{\sim}{N} = \{(k - \lambda)\underset{\sim}{I} + \lambda\underset{\sim}{J}\}\underset{\sim}{N} = \underset{\sim}{N}\{(k - \lambda)\underset{\sim}{I} + \lambda\underset{\sim}{J}\}$,
whence, since $\underset{\sim}{N}$ is not singular,

$$\underset{\sim}{N}'\underset{\sim}{N} = (k - \lambda)\underset{\sim}{I} + \lambda\underset{\sim}{J},$$

so that each pair of blocks have $\lambda$ treatments in common.

This result enables us to obtain two other BIB designs from a symmetric BIBD.

The method of block section consists of omitting from a symmetric BIBD one of its blocks and all the treatments that appear in that block. The plots that remain form what is sometimes called the residual design. It is a BIBD with

$$t' = t - k,\ b' = t - 1,\ r' = k,\ k' = k - \lambda,\ \lambda' = \lambda.$$

Alternatively we can omit a block and retain in the other blocks only those treatments that appear in the omitted block. This is the method of block intersection. The resulting design is sometimes called the derived design. It is a BIBD with parameters

$$t' = k,\ b' = t - 1,\ r' = k - 1,\ k' = \lambda,\ \lambda' = \lambda - 1.$$

Hall and Connor (1953) proved a partial converse theorem: given any BIBD which has the parameters of a residual design with $\lambda = 1$ or 2 it is possible to augment it to obtain the 'original' symmetric BIBD. This shows us, for example, that the design with $t = 15$, $b = 21$, $r = 7$, $k = 5$, $\lambda = 2$ is impossible, because its existence would imply the existence of the symmetric BIBD with $t = 22$, $k = 7$, $\lambda = 2$. The theorem of Hall and Connor is not, however, true when $\lambda = 3$; this is known because of a counterexample obtained by Bhattacharya (1944). He found a design for $t = 16$, $b = 24$, $r = 9$, $k = 6$, $\lambda = 3$, two of whose blocks have four treatments in common; it cannot therefore be augmented to a symmetric design with $t = 25$, $k = 9$, $\lambda = 3$. Bhattacharya's design is reproduced in John (1971, page 282).

# Chapter Five

## GROUP DIVISIBLE DESIGNS

5.1 Introduction. The requirement that $\lambda = r(k-1)/(t-1)$ be an integer sharply curtails the number of feasible sets of parameters for designs. This is particularly true when $t - 1$ is a prime. Consider for example the case $t = 8$, $k = 3$. We have $7\lambda = 2r$, so that $r$ has to be a multiple of seven. Also $rt = bk$ implies that $r$ is a multiple of three. The smallest possible BIBD thus has $r = 21$, $b = 56$; these conditions are satisfied by the unreduced design, i.e., the set of all 56 triples out of eight treatments.

In practical situations, therefore, we are obliged to modify our demand that $V(\hat{\tau}_h - \hat{\tau}_i)$ be the same for all pairs of treatments. As a first step in this direction, we shall consider in this chapter group divisible designs. This will lead us later into the more general class of designs introduced by Bose and Nair (1939), called partially balanced designs.

Suppose that we have $t = mn$ treatments and that we divide them into $m$ groups of $n$ treatments each. We call treatments in the same group first associates, and treatments in different groups second associates. Consider a design with $b$ blocks of size $k$ and $r = bk/t$ in which every treatment appears $\lambda_1$ times with each of its first associates and $\lambda_2$ times with each of its second associates. This is called a group divisible design; the division of the treatments into groups constitutes a group divisible association scheme. We shall see that $V(\hat{\tau}_h - \hat{\tau}_i)$ takes exactly two values, one when the two treatments are first associates and the other when they are second associates.

If $t = 8$, we may denote the treatments by $0, 1, \ldots, 7$ and divide them into four groups: 0 4, 1 5, 2 6, 3 7. The following design has $b = 8$, $r = k = 3$, $\lambda_1 = 0$, $\lambda_2 = 1$:

0 1 3, 1 2 4, 2 3 5, 3 4 6, 4 5 7, 5 6 0, 6 7 1, 7 0 2.

The inequality $b \geq t$ for BIB designs no longer holds; the equation $r(k-1) = \lambda(t-1)$ is now replaced by

$$r(k-1) = \lambda_1(n-1) + \lambda_2 n(m-1). \tag{5.1.1}$$

5.2 The intrablock analysis. We arrange the treatments in order of groups so that the first $n$ treatments form the first group, the next $n$ treatments form the second group, and so on. The concordance matrix $\underset{\sim}{N}\underset{\sim}{N}'$ now takes a convenient form. It can be partitioned into $m^2$ submatrices, each with $n$ rows and $n$ columns. The submatrices along the main diagonal have $r$ for the diagonal elements and $\lambda_1$ for the off-diagonal elements; we write them as $\underset{\sim}{A}_1 = (r-\lambda_1)\underset{\sim}{I}_n + \lambda_1\underset{\sim}{J}_n$. The other submatrices each have the form $\underset{\sim}{A}_2 = \lambda_2\underset{\sim}{J}_n$.

The intrablock matrix, $\underset{\sim}{C}$, has submatrices $r\underset{\sim}{I} - k^{-1}\underset{\sim}{A}_1 = \{(r(k-1) + \lambda_1)\underset{\sim}{I} - \lambda_1\underset{\sim}{J}\}/k$ along the diagonal and $-\lambda_2\underset{\sim}{J}/k$ off the diagonal. If we take $a = \lambda_2/k$, $\underset{\sim}{\Omega}^{-1}$ has square matrices

$$\underset{\sim}{A}_3 = \{(r(k-1) + \lambda_1)\underset{\sim}{I} + (\lambda_2 - \lambda_1)\underset{\sim}{J}\}/k$$

along the diagonal and zero elements elsewhere.

We shall need the following lemma.

Lemma 5.2.1: Let $\underset{\sim}{P} = p\underset{\sim}{I}_n + q\underset{\sim}{J}_n$, $p \neq 0$. Then, unless $nq = -p$, $\underset{\sim}{P}$ is nonsingular and $\underset{\sim}{P}^{-1} = s\underset{\sim}{I}_n + u\underset{\sim}{J}_n$, where $s = p^{-1}$ and $u = -q/(p(p+nq))$.

Proof: Let $(p\underset{\sim}{I}_n + q\underset{\sim}{J}_n)(s\underset{\sim}{I}_n + u\underset{\sim}{J}_n) = \underset{\sim}{I}_n$.
Then $ps\underset{\sim}{I}_n + (pu + qs + nqu)\underset{\sim}{J}_n = \underset{\sim}{I}_n$.
Solving $ps = 1$ and $(pu + qs + nqu) = 0$ gives the desired result.

Applying the lemma to $\underset{\sim}{\Omega}^{-1}$ we see that $\underset{\sim}{\Omega}$ also has square matrices $\underset{\sim}{\Omega}_0$ along the diagonal and zero elements elsewhere:

$$\underset{\sim}{\Omega}_0 = \frac{k}{r(k-1) + \lambda_1}\left[\underset{\sim}{I} - \frac{(\lambda_2 - \lambda_1)}{t\lambda_2}\underset{\sim}{J}\right].$$

It follows that if the $h^{th}$ and $i^{th}$ treatments are first associates,

$$V(\hat{\tau}_h - \hat{\tau}_i) = \frac{2k\sigma^2}{r(k-1) + \lambda_1} = v_1 ;$$

if they are second associates,

$$V(\hat{\tau}_h - \hat{\tau}_i) = \frac{2k\sigma^2}{r(k-1) + \lambda_1}\left[1 - \frac{(\lambda_2 - \lambda_1)}{t\lambda_2}\right] = v_2.$$

We may compare these variances to the variance, $(2\sigma^2/r)$, that would be appropriate with randomized complete blocks, and obtain the two relative effeciencies $E_i = 2/(rv_i)$. For our example with $t = 8$, $k = 3$, we have $v_1 = \sigma^2$, $v_2 = 0.875\sigma^2$, so that $E_1 = 2/3$, $E_2 = 16/21 = 0.76$. Even though $\lambda_1 = 0$, $E_1$ is not very much smaller than $E_2$.

5.3 The latent roots of the concordance matrix. We evaluate the determinant $|\underset{\sim}{N}\underset{\sim}{N}'|$ by applying Lemma 3.2.1 twice:

$$\begin{aligned}|\underset{\sim}{N}\underset{\sim}{N}'| &= |\underset{\sim}{A}_1 - \underset{\sim}{A}_2|^{m-1}|\underset{\sim}{A}_1 + (m-1)\underset{\sim}{A}_2| \\ &= |(r-\lambda_1)\underset{\sim}{I}_n + (\lambda_1 - \lambda_2)\underset{\sim}{J}_n|^{m-1}|(r-\lambda_1)\underset{\sim}{I}_n + (\lambda_1 + (m-1)\lambda_2)\underset{\sim}{J}_n| \\ &= (r-\lambda_1)^{m(n-1)}[r-\lambda_1 + n(\lambda_1 - \lambda_2)]^{m-1}[r + (n-1)\lambda_1 + n(m-1)\lambda_2].\end{aligned}$$

Writing $r - \theta$ for $r$ gives $|\underset{\sim}{N}\underset{\sim}{N}' - \theta\underset{\sim}{I}|$; hence the latent roots of $\underset{\sim}{N}\underset{\sim}{N}'$ are

$$\begin{aligned}\theta_0 &= r + (n-1)\lambda_1 + n(m-1)\lambda_2 = rk, \\ \theta_1 &= r - \lambda_1 + n(\lambda_1 - \lambda_2) = rk - \lambda_2 t, \\ \theta_2 &= r - \lambda_1,\end{aligned}$$

with multiplicities $\alpha_0 = 1$, $\alpha_1 = m-1$, $\alpha_2 = m(n-1)$.

We shall now use these results in two ways, first to establish the nonexistence of some designs and second to classify group divisible designs into three types. In later chapters we shall see how, for the general partially balanced design, the latent roots of $\underset{\sim}{N}\underset{\sim}{N}'$ may be used to calculate $\underset{\sim}{\Omega}$ and the relative efficiencies. The latent roots of the concordance matrix will play a central role in our development of the theory of partially balanced designs.

5.4 Nonexistence of designs. The matrix $\underset{\sim}{N}\underset{\sim}{N}'$ is nonnegative definite, i.e., all its latent roots are nonnegative. This fact enables us to show that for certain sets of parameters which satisfy the conditions that have been given so far in this chapter, no designs exist. Let $m = 2$, $n = 6$, $k = 5$, $r = 10$, $\lambda_1 = 2$, $\lambda_2 = 5$. The condition $r(k-1) = (n-1)\lambda_1 + n(m-1)\lambda_2$ is satisfied, but $\theta_2 = rk - \lambda_2 t = 50 - 60 < 0$.

There is also an extension of Theorem 4.5.1 for symmetric designs.

Theorem 5.4.1: For a symmetric design,

$$|\underset{\sim\sim}{NN'}| = r^2 \theta_1^{(m-1)} \theta_2^{(n-1)m} ,$$

which must be a square. Then, if $\theta_1 \neq 0$ and $\theta_2 \neq 0$, we have two further conditions:

(i) if $m$ is even, $\theta_1$ must be a square;

(ii) if $m$ is odd and $n$ is even, $\theta_2$ must be a square.

We give two examples:

(i) the set of parameters $t = 20$, $m = 2$, $n = 10$, $r = k = 8$, $\lambda_1 = 4$, $\lambda_2 = 2$ has $\theta_1 = 64 - 40 = 24$, which is not a square;

(ii) the set of parameters $t = 54$, $m = 27$, $n = 2$, $r = k = 13$, $\lambda_1 = 0$, $\lambda_2 = 3$ has $\theta_2 = 13 - 0 = 13$, which is not a square.

5.5 The classification of group divisible designs. Bose and Connor (1952) divided the group divisible designs into three classes according to the nature of the latent roots of $\underset{\sim\sim}{NN'}$. A group divisible design is said to be

(i) singular if $\theta_2 = 0$, i.e., $r = \lambda_1$, (5.5.1)

(ii) semiregular if $\theta_1 = 0$, i.e., $rk = \lambda_2 t$, (5.5.2)

(iii) regular if $\theta_1 > 0$ and $\theta_2 > 0$. (5.5.3)

The following design for $t = 6$, $m = 3$, $n = 2$ is singular: 0 3 1 4, 1 4 2 5, 2 5 3 0. Every time a treatment appears its first associate is in the same block. Let D be a BIBD for $m$ treatments; if each of those treatments is replaced by a group of new treatments, the new design is singular group divisible. All singular group divisible designs can be obtained in this way.

Bose and Connor (1952) also proved the following theorem for semiregular designs.

Theorem 5.5.1: A group divisible design is semiregular if, and only if, each block contains $c = k/m$ treatments from each group. Clearly $c$ must be an integer.

Proof: (i) Sufficiency. Every treatment appears in $r$ blocks, each time with $c$ members of each other group. Then $rc = \lambda_2 n$, so that $rk = rcm = \lambda_2 mn = \lambda_2 t$, and the design is thus semiregular.

(ii) Necessity. Suppose that the design is semiregular. We shall show that each block contains $c$ treatments from the first group. Let the $j^{th}$ block contain $x_j$ treatments from the first group. Then $\Sigma x_j = rn$. Each pair of treatments from the group appears $\lambda_1$ times; hence $\Sigma x_j(x_j - 1) = \lambda_1 n(n-1)$. Thus, recalling (5.1.1) and (5.5.2), we have

$$\Sigma x_j^2 = \lambda_1 n(n-1) + rn = n\{rk - n(m-1)\lambda_2\} = n^2\lambda_2.$$

Let $\bar{x} = rn/b = k/m = c$. Then $b\bar{x}^2 = rnc$, and

$$\Sigma(x_j - \bar{x})^2 = n^2\lambda_2 - b\bar{x}^2 = n(n\lambda_2 - rc) = m^{-1}n(\lambda_2 t - rk) = 0,$$

which implies that $x_j = c$ for all $j$.

If we have a semiregular design for $m$ groups, we can clearly strike out the treatments belonging to one or more groups, and still have a semiregular design in the remaining groups.

If one of the replicates is omitted from an OS1 balanced incomplete block design, the remaining blocks form a resolvable semiregular design with $t = n^2$, $m = k = n$, $c = 1$, $\lambda_1 = 0$ and $\lambda_2 = 1$.

Since $\theta_2 > 0$ and $\theta_1 = 0$ the rank of $\underset{\sim\sim}{NN'}$ is $r(\underset{\sim\sim}{NN'}) = \alpha_2 + 1 = t - m + 1$. This leads to an inequality for semiregular designs analagous to Fisher's inequality for BIB designs, since $r(\underset{\sim\sim}{NN'}) \leq b$, namely,

$$b \geq t - m + 1.$$

The following semiregular design has $t = 8$, $m = 4$, $n = 2$ with groups 0 4; 1 5; 2 6; 3 7. There are eight blocks, and $r = k = 4$, $\lambda_1 = 0$, $\lambda_2 = 2$:

0 1 2 3, 0 1 6 7, 0 5 2 7, 0 5 6 3,
4 1 2 3, 4 1 6 7, 4 5 2 7, 4 5 6 3.

If we strike out treatments 3 and 7 we obtain a design for $t = 6$, $m = 3$, $n = 2$, $\lambda_1 = 0$, $\lambda_2 = 2$ with eight distinct blocks. If, on the other hand, we retain 3 and 7, but strike out 0 and 4, we have a design with the same parameters consisting of only four distinct blocks each of which is duplicated.

5.6 The construction of group divisible designs. We shall discuss two standard methods of construction: (i) adding and subtracting replicates from BIB designs; (ii) developing initial blocks cyclically. The basic reference is Bose, Shrikhande, and Bhattacharya (1953).

(i) We have already remarked that omitting a replicate from a BIBD in the OS1 series leaves a semiregular design. We can generalize this idea.

Let D be a BIBD which contains one set of blocks that form a replicate; those blocks divide the treatments into groups. The single replicate is thus a group design with $\lambda_1 = 1$, $\lambda_2 = 0$. If we now take all the blocks that are not in the replicate $a_1$ times each and the blocks that are in the replicate $a_2$ times each, the resulting design is group divisible with $\lambda_1' = a_1(\lambda - 1) + a_2$ and $\lambda_2 = a_1\lambda$.

A variant of this idea is to add to a group divisible design a BIBD with the same treatments and the same block size. The new design has $\lambda_1' = \lambda_1 + \lambda$, $\lambda_2' = \lambda_2 + \lambda$.

(ii) Bose et al. (1953) give an analogue of Bose's fundamental theorem on difference sets for BIB designs. We present a modified version. We denote the treatments by the integers $0, 1, 2, \ldots, t-1 \bmod t$, and group them so that the $i^{th}$ group, $i = 0, 1, \ldots, m-1$ consists of $i, i+n, i+2n, \ldots, i+(m-1)n$. Let A be an initial block such that the $k(k-1)$ differences between its elements consists of each of the first associates of 0 (the multiples of n) exactly $\lambda_1$ times, and each of the other nonzero treatments $\lambda_2$ times. Then developing A cyclically yields a group divisible design with parameters $\lambda_1$ and $\lambda_2$. The procedure can readily be extended to the case of several initial blocks.

Alternatively, we can use two coordinates to represent each treatment with xy denoting the $y^{th}$ treatment in the $x^{th}$ group; $x = 0, 1, \ldots, m-1$; $y = 0, 1, \ldots, n-1$. This is similar to the procedure used in the previous chapter to derive a BIBD for $t = 15$. Addition is defined by $(x_1, y_1) + (x_2, y_2) = (x_1 + x_2, y_1 + y_2)$, where $x_1 + x_2$ is reduced $\bmod m$ and $y_1 + y_2$ is reduced $\bmod n$. We require the differences in the initial block to consist of every $i^{th}$ associate of 00 exactly $\lambda_i$ times.

We illustrate these procedures by three examples.

Example 5.6.1: Let $t = 8$, $m = 4$, $n = 2$. This is the example given in the first section of this chapter. The initial block 0 1 3 gives a design with $\lambda_1 = 0$, $\lambda_2 = 1$.

Example 5.6.2: Let $t = 10$, $m = 5$, $n = 2$. Denote the treatments by 00, 01, 10, 11,... . The following two initial blocks lead to a design with $\lambda_1 = 0$ and $\lambda_2 = 3$:

00 11 21 41; 00 21 31 41.

Example 5.6.3: Again we have $t = 8$. This time we represent the treatments by the elements of the Galois field of $2^3$ elements, i.e., by triples $x_1$, $x_2$, $x_3$ where $x_i = 0$ or 1 and arithmetic is mod 2. The treatments are combined into four groups 000, 111; 001, 110; 010, 101; 100, 011. The initial block 000 010 001 100 provides a set of differences which contains every element other than 000 and 111 exactly twice. The design is

| | | | | | | | |
|---|---|---|---|---|---|---|---|
| 000 | 010 | 001 | 100; | 111 | 101 | 110 | 011; |
| 001 | 011 | 000 | 101; | 110 | 100 | 111 | 010; |
| 010 | 000 | 011 | 110; | 101 | 111 | 100 | 001; |
| 011 | 001 | 010 | 111; | 100 | 110 | 101 | 000. |

The design is semiregular and resolvable. The two blocks in any row form a replicate. Therefore, it is not isomorphic to the design with the same parameters given in Section 5.5, which is not resolvable.

5.7 A further example of combining designs. We give now an example of how group divisible designs with $t = 6$, $k = 3$ can be combined to obtain BIB designs. The treatments are divided into three groups: 0 3; 1 4; 2 5.

The following design, $D_1$, is semiregular with $\lambda_1 = 0$, $\lambda_2 = 2$, (either row has $\lambda_1 = 0$, $\lambda_2 = 1$):

| | | | |
|---|---|---|---|
| 0 1 2 | 2 3 4 | 4 5 0 | 5 1 3 |
| 1 2 3 | 3 4 5 | 5 0 1 | 0 2 4. |

The design can be derived by developing the two initial blocks 0 1 2 and 0 2 4 with a slight modification. When 0 2 4 is developed, the two blocks 0 2 4 and 1 3 5 appear three times each. We take them only once. J. A. John (1966) has called this a partial (in the present case 1/3) cycle.

Designs $D_2$ and $D_3$ are obtained by developing the initial blocks 0 1 3 and 0 1 4 respectively.

$D_2$: 0 1 3, 1 2 4, 2 3 5, 3 4 0, 4 5 1, 5 0 2;

$D_3$: 0 1 4, 1 2 5, 2 3 0, 3 4 1, 4 5 2, 5 0 3.

Each has $\lambda_1 = 2$, $\lambda_2 = 1$.

Between them, $D_1$, $D_2$ and $D_3$ form a partition of the twenty distinct triples of six treatments. Taken together, they form a BIBD with $\lambda = 4$, or we may take just one of $D_2$ or $D_3$ together with either of the rows of $D_1$ and have a BIBD with $\lambda = 2$. Alternatively, we may combine $D_1$ with either $D_2$ or $D_3$ to get a design with $\lambda_1 = 2$, $\lambda_2 = 3$, and so on.

# Chapter Six

# PARTIALLY BALANCED DESIGNS WITH TWO ASSOCIATE CLASSES

6.1 Introduction. Partially balanced incomplete block designs were introduced by Bose and Nair (1939). The group divisible designs considered in the last chapter fall into this category. We motivate our study of these designs by considering two interesting properties of group divisible designs; not only does $V(\hat{\tau}_h - \hat{\tau}_i)$ take only two values but the particular value taken depends only upon whether the treatments are first or second associates. This says, in essence, that $\underset{\sim}{\Omega}$ and $\underset{\sim\sim}{NN'}$ have the same pattern. There are only three distinct elements in $\underset{\sim}{\Omega}$; $\omega_0$ along the main diagonal, $\omega_1$ in the off-diagonal places of those submatrices that lie along the main diagonal, and $\omega_2$ elsewhere. With our choice of $a = \lambda_2/k$ we obtained $\omega_2 = 0$, but with any nonzero value of $a$ the pattern persists; $\omega_0$ occurs where $r$ occurs in $\underset{\sim\sim}{NN'}$; $\omega_1$ appears where $\lambda_1$ appears; $\omega_2$ is found in the same places as $\lambda_2$.

In a survey paper read to the Royal Statistical Society Pearce (1963) advocated the idea that in some incomplete block designs $\underset{\sim}{\Omega}$ and $\underset{\sim\sim}{NN'}$, the covariance matrix and the concordance matrix, might have the same pattern, so that one might classify designs by looking at their concordance matrices without having to evaluate the generalized inverses. In this section we shall tentatively develop that idea and shall find that it leads us to the partially balanced designs of Bose and Nair.

Suppose that we introduce an association scheme for the treatments such that

(i) for each treatment $n_1$ of the other treatments are its first associates, and

(ii) the remaining $n_2 = t - n_1 - 1$ treatments are its second associates;

(iii) if $\alpha$ is a first associate of $\beta$, then $\beta$ is a first associate of $\alpha$.

We consider a proper, binary, equireplicate design in which every treatment appears $\lambda_i$ times with each of its $i^{th}$ associates.

We now introduce two symmetric matrices, $\underset{\sim}{B}_1$ and $\underset{\sim}{B}_2$, each with $t$ rows and $t$ columns. The element in the $h^{th}$ row and the $i^{th}$ column of $\underset{\sim}{B}_j$ is unity if the $h^{th}$ and $i^{th}$ treatments are $j^{th}$ associates, and zero otherwise. It is convenient to think of each treatment as its own zeroth associate, and to define $\underset{\sim}{B}_0 = \underset{\sim}{I}$; $\underset{\sim}{B}_0$, $\underset{\sim}{B}_1$ and $\underset{\sim}{B}_2$ are called association matrices. It follows that

(i) $\underset{\sim}{B}_0 + \underset{\sim}{B}_1 + \underset{\sim}{B}_2 = \underset{\sim}{J}$,

(ii) $\underset{\sim}{B}_0$, $\underset{\sim}{B}_1$, $\underset{\sim}{B}_2$ are linearly independent in the sense that
$c_0\underset{\sim}{B}_0 + c_1\underset{\sim}{B}_1 + c_2\underset{\sim}{B}_2 = \underset{\sim}{0}$ if and only if $c_0 = c_1 = c_2 = 0$,

(iii) $\underset{\sim}{B}_1\underset{\sim}{1} = n_1\underset{\sim}{1}$, $\underset{\sim}{B}_2\underset{\sim}{1} = n_2\underset{\sim}{1}$.

Then

$$\underset{\sim}{N}\underset{\sim}{N}' = r\underset{\sim}{I} + \lambda_1\underset{\sim}{B}_1 + \lambda_2\underset{\sim}{B}_2,$$
$$k\underset{\sim}{C} = r(k-1)\underset{\sim}{I} - \lambda_1\underset{\sim}{B}_1 - \lambda_2\underset{\sim}{B}_2.$$

We should like our design to have the property

$$\underset{\sim}{\Omega} = c_0\underset{\sim}{I} + c_1\underset{\sim}{B}_1 + c_2\underset{\sim}{B}_2.$$

What does this require of $\underset{\sim}{B}_1$ and $\underset{\sim}{B}_2$? We note that

$$\underset{\sim}{B}_1^2 = \underset{\sim}{B}_1(\underset{\sim}{J} - \underset{\sim}{I} - \underset{\sim}{B}_2) = n_1\underset{\sim}{J} - \underset{\sim}{B}_1 - \underset{\sim}{B}_1\underset{\sim}{B}_2$$
$$= n_1\underset{\sim}{I} + (n_1 - 1)\underset{\sim}{B}_1 + n_1\underset{\sim}{B}_2 - \underset{\sim}{B}_1\underset{\sim}{B}_2 = (\underset{\sim}{J} - \underset{\sim}{I} - \underset{\sim}{B}_2)\underset{\sim}{B}_1,$$

and we see that $\underset{\sim}{B}_1$ and $\underset{\sim}{B}_2$ commute; also

$$\underset{\sim}{B}_2^2 = n_2\underset{\sim}{I} + n_2\underset{\sim}{B}_1 + (n_2 - 1)\underset{\sim}{B}_2 + \underset{\sim}{B}_1\underset{\sim}{B}_2.$$

Therefore, if we write $(\underset{\sim}{C} + a\underset{\sim}{J})(c_0\underset{\sim}{I} + c_1\underset{\sim}{B}_1 + c_2\underset{\sim}{B}_2) = \underset{\sim}{I}$, multiply out the left side, express $\underset{\sim}{C}$ and $\underset{\sim}{J}$ in terms of $\underset{\sim}{I}$, $\underset{\sim}{B}_1$, $\underset{\sim}{B}_2$, and simplify, we shall have an equation which can be written as

$$d_0\underset{\sim}{I} + d_1\underset{\sim}{B}_1 + d_2\underset{\sim}{B}_2 + d_{12}\underset{\sim}{B}_1\underset{\sim}{B}_2 = \underset{\sim}{I},$$

where $d_0$, $d_1$, $d_2$, $d_{12}$ are scalars, which are linear combinations of $c_0$, $c_1$ and $c_2$. When $d_{12} \neq 0$ this implies that $\underset{\sim}{B}_1\underset{\sim}{B}_2$ (and hence $\underset{\sim}{B}_1^2$ and $\underset{\sim}{B}_2^2$) is a linear combination of $\underset{\sim}{B}_0$, $\underset{\sim}{B}_1$, $\underset{\sim}{B}_2$.

We may now write

$$\underset{\sim}{B}_1^2 = b_0\underset{\sim}{I} + b_1\underset{\sim}{B}_1 + b_2\underset{\sim}{B}_2.$$

The implications of the requirement are the following:
(i) the inner product of each row of $\underset{\sim}{B}_1$ with itself is $b_0$ (this is satisfied with $b_0 = n_1$);
(ii) the inner product of two rows of $\underset{\sim}{B}_1$ corresponding to first associates is $b_1$ for all pairs;
(iii) the inner product of two rows corresponding to second associates is $b_2$ for all pairs.
Similar conditions will hold for the rows of $\underset{\sim}{B}_2$. In the next section we shall formalize these ideas in the definition of a partially balanced association scheme.

6.2 Partially balanced association schemes. A partially balanced association scheme with two association classes, as defined by Bose and Shimamoto (1952), is a relationship between the $t$ treatments which satisfies the following conditions:
(i) each treatment is a first associate of exactly $n_1$ other treatments and a second associate of the other $n_2 = t - n_1 - 1$ treatments;
(ii) if $t_h$ is an $i^{th}$ associate of $t_{h'}$, then $t_{h'}$ is an $i^{th}$ associate of $t_h$ (symmetry of the relationship);
(iii) if $t_h$ and $t_{h'}$ are $i^{th}$ associates there are exactly $p^i_{jk}$ treatments which are $j^{th}$ associates of $t_h$ and $k^{th}$ associates of $t_{h'}$ where $p^i_{jk}$ is independent of the pair of $i^{th}$ associates chosen (it follows immediately that $p^i_{jk} = p^i_{kj}$).
We consider each treatment to be its own zeroth associate. We define two more symmetric matrices:

$$\underset{\sim}{P}_1 = \begin{bmatrix} p^1_{11} & p^1_{12} \\ p^1_{21} & p^1_{22} \end{bmatrix}; \qquad \underset{\sim}{P}_2 = \begin{bmatrix} p^2_{11} & p^2_{12} \\ p^2_{21} & p^2_{22} \end{bmatrix}.$$

If, in a proper binary equireplicate design, the treatments are associated in a partially balanced scheme and each treatment appears exactly $\lambda_i$ times with each of its $i^{th}$ associates, it is called a partially balanced design with two associate classes [sometimes written PBIB(2) design].

Condition (iii) gives the design the property discussed at the end of the previous section:

$$\underset{\sim}{B}_1^2 = n_1\underset{\sim}{I} + p_{11}^1\underset{\sim}{B}_1 + p_{11}^2\underset{\sim}{B}_2; \qquad \underset{\sim}{B}_2^2 = n_2\underset{\sim}{I} + p_{22}^1\underset{\sim}{B}_1 + p_{22}^2\underset{\sim}{B}_2;$$
$$\underset{\sim}{B}_1\underset{\sim}{B}_2 = p_{12}^1\underset{\sim}{B}_1 + p_{12}^2\underset{\sim}{B}_2. \tag{6.2.1}$$

It follows immediately from the definition that

$$r(k-1) = n_1\lambda_1 + n_2\lambda_2,$$
$$p_{11}^1 + p_{12}^1 = n_1 - 1, \qquad p_{12}^1 + p_{22}^1 = n_2, \tag{6.2.2}$$
$$p_{11}^2 + p_{12}^2 = n_1, \qquad p_{12}^2 + p_{22}^2 = n_2 - 1.$$

The following relationship is less obvious.

<u>Theorem 6.2</u>: $n_1p_{2k}^1 = n_2p_{1k}^2$ for $k = 1$ or $k = 2$.

<u>Proof</u>: Consider any treatment. Let $G_1$ denote the set of its first associates and $G_2$ the set of its second associates. Each treatment in $G_1$ has $p_{2k}^1$ $k^{th}$ associates in $G_2$; each treatment in $G_2$ has $p_{1k}^2$ $k^{th}$ associates in $G_1$. The number of pairs of $k^{th}$ associates, one from $G_1$ and one from $G_2$, is, on the one hand, $n_1p_{2k}^1$ and, on the other hand, $n_2p_{1k}^2$.

6.3 <u>Some association schemes</u>. The three most important partially balanced association schemes with two associate classes are the group divisible, triangular, and latin square schemes.

6.3.1 <u>The group divisible scheme</u>. This was the topic of Chapter 5. The $\underset{\sim}{P}$ matrices are

$$\underset{\sim}{P}_1 = \begin{bmatrix} n-2 & 0 \\ 0 & n(m-1) \end{bmatrix} \quad \text{and} \quad \underset{\sim}{P}_2 = \begin{bmatrix} 0 & n-1 \\ n-1 & n(m-2) \end{bmatrix} ;$$

$n_1 = n - 1$, $n_2 = n(m - 1)$.

We can show that any partially balanced association scheme with $p^1_{12} = 0$, or $p^2_{12} = 0$, must be a group divisible scheme.

6.3.2 The triangular scheme. This is a scheme for $t = n(n-1)/2$ treatments with $n > 4$. To obtain the scheme we construct a symmetric array of $n$ rows and $n$ columns by leaving the main diagonal empty, and writing the treatments in the places above the diagonal (repeating them symmetrically below the diagonal). Two treatments are first associates if they appear in the same row (or the same column) of the array. In the case $n = 5$, $t = 10$ the array is

| | | | | |
|---|---|---|---|---|
| - | 0 | 1 | 2 | 3 |
| 0 | - | 4 | 5 | 6 |
| 1 | 4 | - | 7 | 8 |
| 2 | 5 | 7 | - | 9 |
| 3 | 6 | 8 | 9 | - |

The first associates of 0 are 1, 2, 3, 4, 5, 6; the second associates are 7, 8, 9.

In the general case $n_1 = 2(n-2)$ and $n_2 = (n-2)(n-3)/2$; the $\underset{\sim}{P}$ matrices are

$$\underset{\sim}{P}_1 = \begin{bmatrix} n-2 & n-3 \\ n-3 & (n-3)(n-4)/2 \end{bmatrix}, \quad \underset{\sim}{P}_2 = \begin{bmatrix} 4 & 2(n-4) \\ 2(n-4) & (n-4)(n-5)/2 \end{bmatrix}.$$

In the case $n = 4$ the scheme becomes a group divisible scheme with the classes reversed and $m = 3$, $n = 2$.

An example of a design with the triangular scheme is obtained by considering the rows of the array as blocks. In this design $b = n$, $r = 2$, $k = n$, $\lambda_1 = 1$, $\lambda_2 = 0$.

It has been shown that, if $n \neq 8$, any partially balanced association scheme with two associate classes which has the $\underset{\sim}{P}$ matrices given above must be a triangular scheme. This result was proved for $n \geq 9$ by Connor (1958), for $n \leq 6$ by Shrikhande (1959), and for $n = 7$ by Hoffman (1960). Hoffman also showed that the result was false for $n = 8$. Chang (1960) showed that when $n = 8$ there are three schemes in addition to the triangular scheme which have the same $\underset{\sim}{P}$ matrices (see also Seiden, 1966). Hoffman's proof of the uniqueness of the scheme for $n \neq 8$, and the three pseudotriangular schemes for $n = 8$ may be found in Raghavarao (1971, pp 146 et seq.).

6.3.3 The latin square schemes. These are schemes for $t = n^2$ treatments. The treatments are arranged in a square array. In the simplest of the schemes, the $L_2$ scheme, two treatments are first associates if they appear in the same row or in the same column of the array; otherwise they are second associates. For the $L_i$ scheme with $i > 2$, we superimpose $(i - 2)$ mutually orthogonal latin squares on the array; two treatments are then first associates if they are in the same row, or in the same column, or if they correspond to the same letter in one of the latin squares.

For the $L_i$ scheme, $i \geq 2$, $n_1 = i(n - 1)$ and $n_2 = (n - i + 1)(n - 1)$. The $\underset{\sim}{P}$ matrices are

$$\underset{\sim}{P}_1 = \begin{bmatrix} i^2 - 3i + n & (i-1)(n-i+1) \\ (i-1)(n-i+1) & (n-i)(n-i+1) \end{bmatrix},$$

$$\underset{\sim}{P}_2 = \begin{bmatrix} i(i-1) & i(n-i) \\ i(n-i) & (n-i)^2 + (i-2) \end{bmatrix}.$$

The first $i$ replicates of a BIBD of the OS1 series for $t = n^2$ constitute a $L_i$ design with $\lambda_1 = 1$, $\lambda_2 = 0$. It is not, of course, necessary that we have a complete set of MOLS; $L_2$ and $L_3$ schemes exist for all values of $n > 2$. The $L_2$ design which has just been mentioned for $t = n^2$ consisting of two replicates (actually the array itself together with its transpose) is sometimes called the simple lattice and is commonly used in agricultural experiments. It is particularly useful when $t$ is large.

Shrikhande (1959) showed that the $L_2$ scheme is unique for $n \neq 4$; we shall discuss this special case later. Clatworthy (1973, p.20) notes that whereas the $L_3$ scheme is unique for $n = 3$, it is not unique for $n = 4$, 5 or 6.

6.4 Latent roots and variances. We have already seen that $n_1$ is a latent root of $\underset{\sim}{B}_1$ with the vector $\underset{\sim}{1}$. Let $\varepsilon$ be a root of $\underset{\sim}{B}_1$ which has a vector $\underset{\sim}{x}$ such that $\underset{\sim}{1}'\underset{\sim}{x} = 0$; this will occur if $\varepsilon \neq n_1$, or if the multiplicity of $n_1$ is greater than one. Then

$$\underset{\sim}{B}_2\underset{\sim}{x} = (\underset{\sim}{J} - \underset{\sim}{I} - \underset{\sim}{B}_1)\underset{\sim}{x} = (-1 - \varepsilon)\underset{\sim}{x}.$$

It follows that for the same vector $\underset{\sim}{x}$

$$\Theta = r + \varepsilon\ \lambda_1 - (1 + \varepsilon)\lambda_2$$

is a root of $\underset{\sim}{N}\underset{\sim}{N}'$ with the same multiplicity as $\varepsilon$. Furthermore $\psi = (rk - \Theta)/k$ is a latent root of $\underset{\sim}{C}$ and $\psi^{-1}$ is a root of $\underset{\sim}{\Omega}$ with $\underset{\sim}{x}$ as a vector. The multiplicities of $\varepsilon$, $\Theta$ and $\psi$ are the same.

6.4.1 $\underset{\sim}{B}_1$ <u>has at most three distinct latent roots.</u> We recall from (6.2.1) that

$$\underset{\sim}{B}_1^2 = n_1\underset{\sim}{I} + p_{11}^1\underset{\sim}{B}_1 + p_{11}^2\underset{\sim}{B}_2;\ \underset{\sim}{B}_1\underset{\sim}{B}_2 = p_{12}^1\underset{\sim}{B}_1 + p_{12}^2\underset{\sim}{B}_2.$$

Then
$$\begin{aligned}\underset{\sim}{B}_1^3 &= n_1\underset{\sim}{B}_1 + p_{11}^1\underset{\sim}{B}_1^2 + p_{11}^2 p_{12}^1\underset{\sim}{B}_1 + p_{11}^2 p_{12}^2\underset{\sim}{B}_2 \\ &= n_1\underset{\sim}{B}_1 + p_{11}^1\underset{\sim}{B}_1^2 + p_{11}^2 p_{12}^1\underset{\sim}{B}_1 + p_{12}^2(\underset{\sim}{B}_1^2 - n_1\underset{\sim}{I} - p_{11}^1\underset{\sim}{B}_1),\end{aligned}$$

and we see that $\underset{\sim}{B}_1$ is a root of a cubic equation (unless $p_{11}^2 = 0$). Thus the minimal polynomial of $\underset{\sim}{B}_1$ has degree three or less, so that $\underset{\sim}{B}_1$ has at most three distinct latent roots. We need to say at most three because if $p_{11}^2 = 0$, which is the case for group divisible schemes, the minimal polynomial of $\underset{\sim}{B}_1$ is a quadratic. Substituting $\varepsilon$ for $\underset{\sim}{B}_1$ in the cubic and gathering terms we have

$$\varepsilon^3 - \varepsilon^2(p_{11}^1 + p_{12}^2) + \varepsilon(p_{11}^1 p_{12}^2 - p_{12}^1 p_{11}^2 - n_1) + n_1 p_{12}^2 = 0.$$

We already know that $n_1$ is a root. Factoring out $(\varepsilon - n_1)$, the other two roots are seen to be the roots of the quadratic equation

$$\varepsilon^2 + \varepsilon(n_1 - p_{11}^1 - p_{12}^2) - p_{12}^2 = 0,$$

which is more conveniently written as

$$\varepsilon^2 + (p_{11}^2 - p_{11}^1)\varepsilon - p_{12}^2 = 0.$$

For the triangular scheme this equation becomes

$$\varepsilon^2 - (n-6)\varepsilon - 2(n-4) = 0,$$

with roots $\varepsilon_1 = n-4$, $\varepsilon_2 = -2$.
For the $L_i$ scheme

$$\varepsilon^2 - (n-2i)\varepsilon + i(i-n) = 0$$

with $\varepsilon_1 = n-i$, $\varepsilon_2 = -i$.
For the group divisible scheme with $p_{11}^2 = 0$ we have

$$\underset{\sim}{B}_1^2 = n_1\underset{\sim}{I} + p_{11}^1\underset{\sim}{B}_1; \qquad \varepsilon^2 - (n-2)\varepsilon - (n-1) = 0;$$

the roots are $\varepsilon_1 = n-1 = n_1$, $\varepsilon_2 = -1$.

6.4.2 Multiplicities. The concordance matrix $\underset{\sim}{N}\underset{\sim}{N}'$ has three roots $\theta_0$, $\theta_1$, $\theta_2$; we denote the corresponding roots of $\underset{\sim}{C}$ by $\psi_0$, $\psi_1$, $\psi_2$.

$$\theta_1 = r + \varepsilon_1\lambda_1 - (\varepsilon_1+1)\lambda_2,$$
$$\theta_2 = r + \varepsilon_2\lambda_1 - (\varepsilon_2+1)\lambda_2.$$

If the rank of $\underset{\sim}{C}$ is to be $t-1$, the root $\theta_0 = rk$, which corresponds to $\psi_0 = 0$, must be a simple root; we denote the multiplicities of the other roots by $\alpha_1$, $\alpha_2$ respectively. The trace of $\underset{\sim}{N}\underset{\sim}{N}'$ is $rt$, so that

$$rt = rk + \theta_1\alpha_1 + \theta_2\alpha_2,$$

whence

$$r(k-1) = n_1\lambda_1 + n_2\lambda_2 = -\lambda_1(\alpha_1\varepsilon_1 + \alpha_2\varepsilon_2) + \lambda_2(\alpha_1 + \alpha_1\varepsilon_1 + \alpha_2 + \alpha_2\varepsilon_2).$$

The multiplicities are the multiplicities of roots of the association matrices and are independent of $\lambda_1$, $\lambda_2$. The last equation is thus an identity for $\lambda_1$, $\lambda_2$, so that we have two equations to solve for $\alpha_i$:

$$\alpha_1\varepsilon_1 + \alpha_2\varepsilon_2 = -n_1, \qquad \alpha_1 + \alpha_2 = t-1 = n_1 + n_2,$$

which have the solutions

$$\alpha_1(\varepsilon_1 - \varepsilon_2) = -n_1(1 + \varepsilon_2) - n_2\varepsilon_2,$$
$$\alpha_2(\varepsilon_1 - \varepsilon_2) = n_1(1 + \varepsilon_1) + n_2\varepsilon_1.$$

We have already obtained $\alpha_1$ and $\alpha_2$ for the group divisible scheme by a different method. We can now show that for the triangular scheme

$$\theta_1 = r + (n - 4)\lambda_1 - (n - 3)\lambda_2, \qquad \alpha_1 = n - 1,$$
$$\theta_2 = r - 2\lambda_1 + \lambda_2, \qquad \alpha_2 = n(n - 3)/2,$$

and for the latin square schemes

$$\theta_1 = r - (i - n)(\lambda_1 - \lambda_2) - \lambda_2, \qquad \alpha_1 = i(n - 1),$$
$$\theta_2 = r - i(\lambda_1 - \lambda_2) - \lambda_2, \qquad \alpha_2 = (n - 1)(n - i + 1).$$

6.4.3 Efficiencies. Suppose that we write

$$\underset{\sim}{\Omega} = \omega_0\underset{\sim}{I} + \omega_1\underset{\sim}{B}_1 + \omega_2\underset{\sim}{B}_2;$$

then, if varieties $h$ and $h'$ are $i^{th}$ associates,

$$V(\hat{\tau}_h - \hat{\tau}_{h'}) = 2\sigma^2(\omega_0 - \omega_i), \qquad E_i^{-1} = r(\omega_0 - \omega_i).$$

The latent roots of $\underset{\sim}{\Omega}$ are obtained from the latent roots of $\underset{\sim}{N}\underset{\sim}{N}'$ by writing $\omega_0$ for $r$, $\omega_1$ for $\lambda_1$ and $\omega_2$ for $\lambda_2$, so that $\omega_0 - \omega_1$ and $\omega_0 - \omega_2$ are the solutions to the two equations:

$$\psi_1^{-1} = \omega_0 + \varepsilon_1\omega_1 - (1 + \varepsilon_1)\omega_2$$
$$= -(\omega_0 - \omega_1)\varepsilon_1 + (\omega_0 - \omega_2)(1 + \varepsilon_1);$$
$$\psi_2^{-1} = -(\omega_0 - \omega_1)\varepsilon_2 + (\omega_0 - \omega_2)(1 + \varepsilon_2).$$

It follows that

$$(\omega_0 - \omega_1)(\varepsilon_1 - \varepsilon_2) = (1 + \varepsilon_1)\psi_2^{-1} - (1 + \varepsilon_2)\psi_1^{-1},$$
$$(\omega_0 - \omega_2)(\varepsilon_1 - \varepsilon_2) = \varepsilon_1\,\psi_2^{-1} - \varepsilon_2\,\psi_1^{-1}.$$

6.5 Intrablock analysis. Since $\hat{\tau} = \Omega Q$, we may write

$$\hat{\tau}_h = \omega_0 Q_h + \omega_1 S_1(Q_h) + \omega_2 S_2(Q_h)$$

where $S_i(Q_h)$ denotes the sum $\Sigma Q_{h'}$, taken over all the treatments $h'$ that are $i^{th}$ associates of $h$. Subtracting

$$0 = \omega_2\{Q_h + S_1(Q_h) + S_2(Q_h)\}$$

from this equation, we have the more convenient form

$$\hat{\tau}_h = (\omega_0 - \omega_2)Q_h + (\omega_1 - \omega_2)S_1(Q_h).$$

The notation $S_i(Q_h)$ was introduced by C. R. Rao (1947).

Clatworthy (1973) includes with each design in his tables two numbers $c_1$ and $c_2$. These numbers occur in the solution to the normal equations obtained by Bose and Shimamoto (1952). They are related to $(\omega_0 - \omega_1)$ and $(\omega_0 - \omega_2)$ by the equation

$$(\omega_0 - \omega_i)r(k - 1) = k - c_i.$$

Clatworthy also includes two other parameters $H$ and $\Delta$. They are relevant only when the experimenter wishes to combine the estimates from the usual intrablock analysis with the results of another method of analysis called the interblock analysis. The latter form of analysis is of limited application, and we shall not discuss it. The reader who is interested may refer to the original paper by Bose and Shimamoto, to Clatworthy (1973), or to John (1971, Chapter 12).

# Chapter Seven

# THE CONSTRUCTION OF PARTIALLY BALANCED DESIGNS WITH TWO ASSOCIATE CLASSES

7.1 Introduction. The main source of information about two associate-class designs is the set of tables by Clatworthy, which was published by the National Bureau of Standards in 1973. This is a revised and vastly expanded version of the original tables prepared by Bose, Clatworthy and Shrikhande (1954), and contains almost a thousand designs with $r \leq 10$, $k \leq 10$.

7.2 The dual of an incomplete block design. The idea of the dual of a design is based on a concept of geometry. We can sometimes take a theorem about points and lines and, by writing lines where we had said points and vice versa, obtain a new theorem. The new theorem is called the dual of the original theorem.

Suppose that we consider an incomplete block design as a geometric figure with $t$ points (treatments) and $b$ lines (blocks); the lines are not necessarily straight lines in the Euclidean tradition. There are exactly $r$ lines passing through each point, and exactly $k$ points on each line. The dual design has $t$ blocks and $b$ treatments. The statement "the first block contains treatments A, B, C, D" becomes "the first treatment is contained in blocks a, b, c, d." We shall denote the dual of the design by $D$ by $D^*$.

Consider the following design with the triangular scheme and $t = 10$, $b = 5$, $r = 2$, $k = 4$, $\lambda_1 = 1$, $\lambda_2 = 0$. The blocks are the rows of the array:

0 1 2 3, 0 4 5 6, 1 4 7 8, 2 5 7 9, 3 6 8 9.

The dual design is A B, A C, A D, A E, B C, B D, B E, C D, C E, D E, which is the unreduced BIBD with $t = 5$, $k = 2$.

The dual of a BIBD has the property that any two blocks have exactly $\lambda$ treatments in common. Such a design is called a linked block design, and has $t \geq b$. The dual of a symmetric BIBD is itself a symmetric BIBD.

The concordance matrix of $D^*$ is $\underset{\sim}{N}'\underset{\sim}{N}$. The two matrices $\underset{\sim}{N}\underset{\sim}{N}'$ and $\underset{\sim}{N}'\underset{\sim}{N}$ have the same nonzero latent roots with the same multiplicities. They have the same rank, and so, if $D$ is not symmetric, either $D$ or $D^*$ must have a zero latent root whose multiplicity absorbs the difference between $t$ and $b$. The dual of a linked block design has two nonzero roots: $\theta_0 = rk$ is simple; the other root is $k(b-r)/(b-1)$ with multiplicity $b-1$.

It follows that necessary and sufficient conditions for a connected incomplete block design with $t > b$ to be a linked block design are that

(i)  $\underset{\sim}{N}\underset{\sim}{N}'$ shall have three distinct latent roots, one of which is the simple root $rk$,

(ii)  one of the other two roots shall be $\theta = k(b-r)/(b-1)$ with $\alpha = b-1$,

(iii)  the other root shall be zero with multiplicity $t-b$.

We have just seen an example of a triangular linked block design; another will be given later in this chapter. John (1970) has shown there are no linked block designs with the $L_i$ scheme for $i = 2, 3, 4$. There are two possibilities for linked block group divisible designs. If $\theta_2 = 0$, we have singular designs with $b - 1 = \alpha = m - 1$. These are obtained by taking a symmetric BIBD for $m$ treatments, and replacing each treatment by $n$ new treatments. If $\theta_1 = 0$ we have semiregular designs with $b = t - m - 1$. The duals of the OS1 designs form such a series.

7.3  <u>Some standard constructions.</u>  The two methods of construction presented in this section are applicable to all the partially balanced association schemes, including those with more than two associate classes that will be discussed later.

(i)  The elementary symmetric designs.

These are two series of designs. In the ES(i) design the $j^{th}$ block consists of all the $i^{th}$ associates of the $j^{th}$ treatment. Then $r = k = n_i$, $\lambda_i = p_{ii}^i$, $\lambda_h = p_{ii}^h$ $(h \neq i)$. In the E'S(i) design the $j^{th}$ block consists of all of the $i^{th}$ associates of the $j^{th}$ treatment together with the $j^{th}$ treatment itself; $r = k = n_i + 1$, $\lambda_i = p_{ii}^i + 2$, $\lambda_h = p_{ii}^h$. The ES(1) and E'S(2) designs are complementary designs; so are the ES(2) and E'S(1) designs.

These designs are not very attractive for group divisible schemes because so many blocks are duplicates. Only the E'S(2) design is disconnected. The ES(1) and E'S(2) designs for the triangular scheme with

$n = 6$ are BIB designs for $t = 15$ with $k = 8$, $\lambda = 4$ in the first case and $k = 7$, $\lambda = 3$ in the second. The ES(1) design for the $L_2$ scheme with $t = 16$ is a BIBD with $k = 6$, $\lambda = 2$; the ES(1) design for the $L_3$ scheme with $t = 36$ is a BIBD with $k = 15$, $\lambda = 6$.

(ii) Subsets from the same class.

In all association schemes there are sets of mutual first or second associates. We can utilize this property by taking subsets of these sets as blocks. Clearly we can take the simple lattice and, by replacing each block by a BIBD in the treatments in that block, obtain a new $L_2$ design. For example, the simple lattice for $t = 16$ has eight blocks of size four. Replacing each block by a BIBD for four treatments in four blocks of size three gives us an $L_2$ design with $t = 16$, $b = 32$, $r = 6$, $k = 3$, $\lambda_1 = 2$, $\lambda_2 = 0$.

For the triangular association schemes we can make use of the fact that each row of the basic array constitutes a set of $n - 1$ mutual first associates.

7.4 Some methods of construction for triangular designs. There are several construction techniques which use the particular properties of this association scheme.

(i) We have seen that taking the rows of the array as blocks gives a design with $\lambda_1 = 1$, $\lambda_2 = 0$. These designs are duals of the unreduced BIB designs for $t = n$, $k = 2$, $\lambda = 1$, and are called the singly linked block series.

(ii) Shrikhande (1952) showed that the duals of BIB designs with $t = (n - 1)(n - 2)/2$, $b = n(n - 1)/2$, $r = n$, $k = n - 2$, $\lambda = 2$ are triangular designs with $\lambda_1 = 1$, $\lambda_2 = 2$.

(iii) Each block of the singly linked designs contains a set of $n - 1$ mutual first associates. When there exists a BIBD with $t = n - 1$, $k < n - 1$, we can replace each row by such a BIBD. In particular, we may replace each block by the $n - 1$ subsets of $n - 2$ treatments each and obtain a design with $t = n(n - 1)/2$, $b = n(n - 1)$, $r = 2(n - 2)$, $k = n - 2$, $\lambda_1 = n - 3$, $\lambda_2 = 0$.

(iv) Clatworthy (1956) obtains designs with $k = 3$, $r = n - 2$, $\lambda_1 = 1$, $\lambda_2 = 0$ when $n \geq 5$. He takes triples of mutual first associates. They consist of a pair in the same row together with the treatment which is found at the intersection of the second of the rows in which the first treatment appears and the column in which the second treatment appears.

For the case $t = 10$ the blocks are

1 2 5, 1 3 6, 1 4 7, 2 3 8, 2 4 9, 3 4 0, 5 6 8, 5 7 9, 6 7 0, 8 9 0.

Clatworthy (1973) lists several other procedures. There are also some designs listed which have been obtained by procedures that are specific to that particular set of parameters.

7.5 Designs with latin square schemes. We have already shown how the fact that in an $L_i$ scheme the treatments form sets of $n-1$ mutual first associates may be exploited in constructing designs.

The triangular scheme is not amenable to the derivation of designs by developing initial blocks. That technique will not work for the $L_2$ scheme either if we confine ourselves to representing the treatments by the integers $0, 1, \ldots, t-1$. If, however, we use a two coordinate system, representing the treatment in the $x^{th}$ row and the $y^{th}$ column of the array by the pair $xy$, and carrying out our arithmetic on each coordinate $\bmod n$, the elementary symmetric designs may be obtained by cyclical development. In the case $t=9$ the ES(1) design becomes

| | | | | | | | | | | | |
|---|---|---|---|---|---|---|---|---|---|---|---|
| 01 | 02 | 10 | 20 | 11 | 12 | 20 | 00 | 21 | 22 | 00 | 10 |
| 02 | 00 | 11 | 21 | 12 | 10 | 21 | 01 | 22 | 20 | 01 | 11 |
| 00 | 01 | 12 | 22 | 10 | 11 | 22 | 02 | 20 | 21 | 02 | 12 |

with $k=4$, $\lambda_1=1$, $\lambda_2=2$.

If $n$ is even, $n/2$ repetitions of the simple lattice form a symmetric design. This cannot be obtained by developing a single initial block; however, if we take any row of the array and any column as two initial blocks and develop them, we get the simple lattice $n$ times. A design with $k=n$ and $\lambda_2=0$ must be the simple lattice repeated $\lambda_1$ times.

# Chapter Eight

## OTHER ASSOCIATION SCHEMES WITH TWO CLASSES

8.1 Introduction. In this chapter we discuss three topics: (1) the two latin square association schemes for $t = 16$; (2) the cyclic association schemes; (3) partial geometries.

8.2 Latin square schemes with $t = 16$. The $L_2$ scheme with $t = 16$ has $n_1 = 6$, $n_2 = 9$. The $\underset{\sim}{P}$ matrices are

$$\underset{\sim}{P}_1 = \begin{bmatrix} 2 & 3 \\ 3 & 6 \end{bmatrix}, \qquad \underset{\sim}{P}_2 = \begin{bmatrix} 2 & 4 \\ 4 & 4 \end{bmatrix}.$$

Each row and each column of the square array forms a set of four mutual first associates.

An $L_3$ scheme with $t = 16$ has $n_1 = 9$, $n_2 = 6$. If we relabel first associates as second associates and vice versa, we have an association scheme with the same $\underset{\sim}{P}$ matrices as the $L_2$ scheme.

The latin squares of side four fall into two types. Two typical squares are given below, superimposed on an array of sixteen treatments, numbered 1 through 16.

$$L_3^*(4) = \begin{matrix} 1A & 2B & 3C & 4D \\ 5D & 6A & 7B & 8C \\ 9C & 10D & 11A & 12B \\ 13B & 14C & 15D & 16A \end{matrix}, \qquad L_3(4) = \begin{matrix} 1A & 2B & 3C & 4D \\ 5B & 6A & 7D & 8C \\ 9C & 10D & 11A & 12B \\ 13D & 14C & 15B & 16A \end{matrix}.$$

Any latin square of side four can be transformed by changing letters and permuting rows and columns into one or the other, but not both. The notation $L_3^*(4)$ and $L_3(4)$ is used by Clatworthy (1973). Only the $L_3(4)$

squares provide MOLS.

Suppose now that we take $L_3$ schemes with these squares and relabel the treatments as suggested above to obtain schemes with $L_2$ parameters. In both cases the first associates of 1 are 7 8 12 10 14 15. In the scheme derived from $L_3(4)$, the first associates can be split into triples to form with treatment 1 two sets of mutual first associates 1 7 12 14; 1 8 10 15. We can then rearrange the numbers in the array to form a "traditional" $L_2$ scheme:

| | | | |
|---|---|---|---|
| 1 | 7 | 12 | 14 |
| 8 | 2 | 13 | 11 |
| 10 | 16 | 3 | 5 |
| 15 | 9 | 6 | 4 . |

In the $L_3^*(4)$ case, however, we can find triples of mutual first associates, but no sets of four. Thus, the two schemes are not isomorphic, and the second scheme is called a pseudo-latin square scheme. The triples of first associates for the pseudo-scheme are listed below in rows of eight.

1 7 14; 2 8 15; 3 5 16; 4 6 13; 5 11 2; 6 12 3; 7 9 4; 8 10 1;
9 15 6; 10 16 7; 11 13 8; 12 14 5; 13 3 10; 14 4 11; 15 1 12; 16 2 9;
1 7 10; 2 8 11; 3 5 12; 4 6 9; 5 11 14; 6 12 15; 7 9 16; 8 10 13;
9 15 2; 10 16 3; 11 13 4; 12 14 1; 13 3 6; 14 4 7; 15 1 8; 16 2 5.

Either the first two rows or the last two rows form a design with $k = 3$, $\lambda_1 = 1$, $\lambda_2 = 0$. There is not a design with these parameters for the regular $L_2$ scheme. On the other hand, the simple lattice from the $L_2$ scheme has $k = 4$, $b = 8$, $\lambda_1 = 1$, $\lambda_2 = 0$, and the pseudo-scheme has no design with those parameters.

If we number the rows 0, 1, 2, 3 and the columns 0, 1, 2, 3, and perform arithmetic mod 4, each of the designs for $k = 3$ is obtained by cyclical development. The first six blocks of the first row now become

00 12 31, 01 13 32, 02 10 33, 03 11 30, 10 22 01, 11 23 02.

The difference set 00 12 31 contains the differences 12 - 00 = 12, 31 - 00 = 31, 31 - 12 = 23, 00 - 12 = 32, 00 - 31 = 13, 12 - 31 = 21, which are the first associates of 00, once each.

Similarly, the ES(1) design is obtained by developing the initial block 12 13 21 23 31 32.

The ES(1) design for the $L_2$ scheme derived from the $L_3(4)$ square has for its first three blocks the first associates of 1, 2, and 3. They are

7 8 10 12 14 15, 7 8 9 11 13 16, 5 6 10 12 13 16 .

If the treatments are represented by two coordinates from 0, 1, 2, 3 we do not get the second and third blocks by developing the first block. If, however, we number the rows and columns as 0, 1, x, y (where $y = 1 + x$), which are the elements of $GF(2^2)$, with arithmetic mod 2, the three blocks become

1x 1y x1 xy y1 yx , 1y 1x x0 xx y0 yy , 10 11 xy x1 yy y0.

We can now obtain the second block by adding 01 to each member of the first block, and the third by adding 0x to each member. With this representation all sixteen blocks are obtained by developing the first block.

We mention in passing that Mesner (1967) has obtained negative latin square schemes, $NL_i(n)$, by letting both i and n be negative. Little use has been made of these schemes. The $NL_2(4)$ scheme for sixteen treatments will be mentioned later in connection with the hypercubic association scheme for $2^4$ treatments.

8.3 <u>Cyclic association schemes</u>. Suppose that the treatments are represented by the elements of an abelian group of t elements. In a cyclic association scheme we take a subset $D = (d_1, d_2, \ldots, d_{n_1})$ of nonzero elements of the group. The first associates of treatment h are $(h + d_1, h + d_2, \ldots)$; the other treatments are its second associates. With this scheme the ES(1) design is obtained by developing the initial block D.

If the scheme is to be partially balanced with two associate classes, there are two conditions which must be satisfied. The differences in D, the initial block, have to consist of every member of D the same number, $\alpha$, of times, and every other nonzero treatment $\beta$ times; we shall have $\alpha = p^1_{11} = \lambda_1$ and $\beta = p^2_{11} = \lambda_2$. The second requirement is that if $d_i$ is an element in D, then $-d_i$ is also in D. To see the need for this requirement we note that, for example, if $h + d_i$ is a first associate of h, then

$h$ is a first associate of $h + d_i$, and so there must exist $d_j$ in $D$ such that $(h + d_i) + d_j = h$, which implies $d_j = -d_i$.

We have already seen that group divisible and $L_2$ schemes can be represented in this way, but it is customary to exclude them from this category. When Bose and Shimamoto (1952) introduced these schemes, they confined their attention to the case in which the treatments were represented by $0, 1, \ldots, t-1 \pmod t$. They gave very few examples. In all their examples $t = 4m + 1$ and is a prime, $n_1 = n_2 = 2m$,

$$\underset{\sim}{P}_1 = \begin{bmatrix} m-1 & m \\ m & m \end{bmatrix}, \qquad \underset{\sim}{P}_2 = \begin{bmatrix} m & m \\ m & m-1 \end{bmatrix}.$$

The subset $D$ was either the set of quadratic residues $\pmod t$ i.e., the set of elements which are squares in $GF(4m+1)$ or, equivalently, the set of nonquadratic residues (nonzero nonsquares).

An association scheme for $t = 4m + 1$ which has the $\underset{\sim}{P}$ matrices given above is said to be pseudocyclic, a classification intended to include schemes which have these parameters, but may not be cyclic. Mesner (1965) has shown that if $t$ is prime, then cyclic association schemes exist only if $t = 4m + 1$, and then they must have $\underset{\sim}{P}$ matrices of the pseudocyclic type. Shrikhande and Singh (1962) have shown that such schemes do not exist if the square free part of $4m + 1$ contains a prime which is congruent to $3 \pmod 4$.

The set of quadratic residues for $t = 9$, $GF(3^2)$, gives us a scheme which is equivalent to the $L_2$ scheme. For $t = 25$ the elements of $GF(5^2)$ are written $pq$ $(= px + q)$, where $p$ and $q$ take the values 0, 1, 2, 3, 4. Arithmetic is mod 5 and in multiplication $x^2 = 2$. Taking $y = 1 + 2x$ as a primitive element the quadratic residues form a suitable set $D$. In the order $y^2, y^4, y^6, \ldots,$ they are

44 23 03 22 14 04 11 32 02 33 41 01 .

In the differences each quadratic residue appears $m - 1 = 5$ times; each other element appears $m = 6$ times.

8.4 Partial geometries. In the original listing of designs by Bose, Clatworthy and Shrikhande (1954) there were several designs with $\lambda_1 = 1$, $\lambda_2 = 0$ which belonged to no standard association scheme. They were gathered together under the name of simple designs. Bose (1963) defined partial geometries; these geometries lead to association schemes with two classes, and it was found that all the simple designs that had hitherto been unclassified corresponded to association schemes of this kind.

A partial geometry is a collection of $t$ points and $b$ lines satisfying the following axioms of incidence:

I. there is at most one line joining any two points;

II. there are $r$ lines through each point;

III. there are $k$ points on each line;

IV. if $P$ is a point and $q$ is a line which does not contain $P$, there are exactly $m$ $(\geq 1)$ lines which join $P$ to points on $q$;

It can be shown that if we now take points as treatments and lines as blocks, we have a partially balanced design with two associate classes and parameters:

$$t = m^{-1}k\{(r-1)(k-1)+m\}, \qquad b = m^{-1}r\{(r-1)(k-1)+m\},$$

$$n_1 = r(k-1), \qquad n_2 = m^{-1}(r-1)(k-1)(k-m),$$

$$\lambda_1 = 1, \ \lambda_2 = 0,$$

$$\underset{\sim}{P}_1 = \begin{bmatrix} (m-1)(r-1)+k-2 & (r-1)(k-m) \\ (r-1)(k-m) & (r-1)(k-m)(k-m-1)/m \end{bmatrix},$$

$$\underset{\sim}{P}_2 = \begin{bmatrix} rm & r(k-m-1) \\ r(k-m-1) & \{(r-1)(k-1)(k-2m)+m(rm-k)\}/m \end{bmatrix}.$$

The reader is referred to Clatworthy (1973) for a listing of designs derived from partial geometries. Raghavarao (1971, p. 194) proves that if a design has the above parameters and $k > r$, it is a partial geometry. He also points out that its dual too is a partial geometry.

We may use the results of chapter six to show that, in computing the roots of $\underset{\sim}{N}\underset{\sim}{N}'$, $\varepsilon_1 = -r$ and $\varepsilon_2 = (k-m-1)$. Hence, since the multiplicities,

$\alpha_1$ and $\alpha_2$, must be integers, a necessary condition for the existence of a partial geometry is that

$$\frac{rk(r-1)(k-1)}{m(k+r-m-1)}$$

should be a positive integer.

# Chapter Nine

## PARTIALLY BALANCED ASSOCIATION SCHEMES WITH SEVERAL ASSOCIATE CLASSES

9.1 Introduction. The idea of partially balanced designs with two associate classes can be extended to association schemes and designs with $m$ classes $(m > 2)$, although in practice one is rarely concerned with more than three classes. We shall first define an m-class scheme and give an example with $m = 3$. Then we shall extend the results of chapter six about association matrices, latent roots, and efficiencies.

9.2 Definition of an association scheme. An association scheme for $t$ treatments is said to be partially balanced with $m$ associate classes if it satisfies the following conditions:

(i) Each treatment has among the other treatments exactly $n_i$ $i^{th}$ associates, $1 \le i \le m$, where $n_i > 0$ for each $i$, and $\Sigma n_i = t - 1$.

(ii) Let $t_1$ and $t_2$ be any two treatments; if $t_1$ is an $i^{th}$ associate of $t_2$, then $t_2$ is an $i^{th}$ associate of $t_1$.

(iii) If $t_1$ and $t_2$ are $i^{th}$ associates, then among the other treatments there are exactly $p^i_{jk}$ which are both $j^{th}$ associates of $t_1$ and $k^{th}$ associates of $t_2$; the number $p^i_{jk}$ is independent of our choice of the pair of $i^{th}$ associates; it follows immediately that $p^i_{jk} = p^i_{kj}$.

There are now $m$ $\underset{\sim}{P}$ matrices, each with $m$ rows and $m$ columns. The following relationships between the parameters $p^i_{jk}$ are proved in the same way as their equivalents in chapter six.

$$\sum_{k=1}^{m} p^i_{jk} = n_j - \delta_{ij}, \quad \text{for each } i,$$

where $\delta_{ij}$ is the Kronecker delta, i.e., $\delta_{ij} = 0$ if $i \neq j$ and $\delta_{ij} = 1$ if $i = j$.

We also have for all $i, j, k$

$$n_i p^i_{jk} = n_j p^j_{ik}.$$

We shall mean by a partially balanced design with this scheme a proper binary equireplicate design in which each pair of $i^{th}$ associates appear together in exactly $\lambda_i$ blocks for each $i$. It was at first thought, wrongly, that the $\lambda_i$ had to be all different. This was later found to be unnecessary, although sometimes the equality of two lambdas brings a reduction in the number of associate classes. The following relationship is easily established:

$$r(k-1) = \sum_{i=1}^{m} n_i \lambda_i.$$

We again regard each treatment as its own zeroth associate. We therefore write

$$n_0 = 1, \quad p^0_{ij} = n_i \delta_{ij}, \quad p^i_{0k} = \delta_{ik}, \quad p^0_{00} = 1.$$

9.3 <u>The rectangular association scheme</u>. This is an association scheme with three classes for $t = mn$ which was introduced by Vartak (1955). We arrange the treatments in a rectangular array with $m$ rows and $n$ columns. Two treatments are first associates if they appear in the same row of the array, second associates if they appear in the same column; otherwise they are third associates. We have

$$n_1 = n - 1, \; n_2 = m - 1, \; n_3 = (m-1)(n-1).$$

The $\underset{\sim}{P}$ matrices are

$$\underset{\sim}{P}_1 = \begin{bmatrix} n-2 & 0 & 0 \\ 0 & 0 & m-1 \\ 0 & m-1 & (m-1)(n-2) \end{bmatrix},$$

$$P_2 = \begin{bmatrix} 0 & 0 & n-1 \\ 0 & m-2 & 0 \\ n-1 & 0 & (m-2)(n-1) \end{bmatrix},$$

$$P_3 = \begin{bmatrix} 0 & 1 & n-2 \\ 1 & 0 & m-2 \\ n-2 & m-2 & (m-2)(n-2) \end{bmatrix}.$$

For $t = 12$, $m = 4$, $n = 3$, we may take the array

| | | |
|---|---|---|
| 1 | 2 | 3 |
| 4 | 5 | 6 |
| 7 | 8 | 9 |
| 10 | 11 | 12 . |

The following design has $r = k = 6$, $\lambda_1 = 3$, $\lambda_2 = 4$, $\lambda_3 = 2$. It is the ES(3) design. The $j^{th}$ block consists of all six third associates of the $j^{th}$ treatment, $\lambda_i = p^i_{33}$:

| | | |
|---|---|---|
| 5 6 8 9 11 12 | 4 6 7 9 10 12 | 4 5 7 8 10 11 |
| 2 3 8 9 11 12 | 1 3 7 9 10 12 | 1 2 7 8 10 11 |
| 2 3 5 6 11 12 | 1 3 4 6 10 12 | 1 2 4 5 10 11 |
| 2 3 5 6 8 9 | 1 3 4 6 7 9 | 1 2 4 5 7 8 . |

9.4 Latent roots of the association and concordance matrices. We now have $m + 1$ association matrices $B_0 = I$, $B_1, \ldots, B_m$. In $B_j$ the element in the $h^{th}$ row and the $i^{th}$ column is unity if the $h^{th}$ and $i^{th}$ treatments are $j^{th}$ associates and is zero otherwise. The concordance matrix is

$$NN' = rB_0 + \sum_{j=1}^{m} \lambda_j B_j.$$

The association matrices are linearly independent: $\sum_{j=0}^{m} c_j B_j = 0$ if, and only if, $c_0 = c_1 = \cdots = c_m$. Thus the matrices form the basis for a vector space. Their products are also in the vector space for

$$B_i B_j = \sum_{k=0}^{m} p^k_{ij} B_k = \sum_{k=0}^{m} p^k_{ji} B_k = B_j B_i.$$

In particular

$$\underset{\sim}{B}_i^2 = \sum_{k=0}^{m} p_{ii}^k \underset{\sim}{B}_k = n_i \underset{\sim}{I} + \sum_{k=1}^{m} p_{ii}^k \underset{\sim}{B}_k.$$

Every integral power of $\underset{\sim}{B}_i$ may be written as a linear combination of the $\underset{\sim}{B}_j$. It follows that $\underset{\sim}{B}_i$ must satisfy a polynomial equation of degree no higher than $m+1$. In general, therefore, the minimal polynomial of $\underset{\sim}{B}_i$ has degree $m+1$, or less, and $\underset{\sim}{B}_i$ has no more than $m+1$ distinct latent roots. One of these roots is $n_i$ with corresponding vector $\underset{\sim}{1}$.

The linear combinations of the $\underset{\sim}{B}_i$ form a ring with a unit element, $\underset{\sim}{B}_0$; in particular, $\underset{\sim}{C} + a\underset{\sim}{J}$ is an element of the ring and so is its inverse $\underset{\sim}{\Omega}$. Thus $\underset{\sim}{\Omega}$ is a linear combination of the $\underset{\sim}{B}_i$, and we may write $\underset{\sim}{\Omega} = \omega_0 \underset{\sim}{B}_0 + \omega_1 \underset{\sim}{B}_1 + \cdots + \omega_m \underset{\sim}{B}_m$; $\underset{\sim}{\Omega}$ has the same pattern as $\underset{\sim}{N}\underset{\sim}{N}'$.

The triple product $\underset{\sim}{B}_h \underset{\sim}{B}_s \underset{\sim}{B}_i$ may be calculated in several ways:

$$\underset{\sim}{B}_h(\underset{\sim}{B}_s \underset{\sim}{B}_i) = \sum_k p_{si}^k \underset{\sim}{B}_h \underset{\sim}{B}_k = \sum_{uk}\sum p_{si}^k p_{hk}^u \underset{\sim}{B}_u,$$

$$(\underset{\sim}{B}_h \underset{\sim}{B}_s)\underset{\sim}{B}_i = \sum_{uk}\sum p_{hs}^k p_{ik}^u \underset{\sim}{B}_u,$$

$$(\underset{\sim}{B}_h \underset{\sim}{B}_i)\underset{\sim}{B}_s = \sum_{uk}\sum p_{hi}^k p_{sk}^u \underset{\sim}{B}_u.$$

We thus have, for each value of $u$, $u = 0, 1, \ldots, m$,

$$\sum_k p_{si}^k p_{hk}^u = \sum_k p_{sh}^k p_{ki}^u = \sum_k p_{hi}^k p_{sk}^u.$$

Following Bose and Mesner (1959) we now introduce a collection of matrices $\underset{\sim}{\Pi}_i$, $i = 0, 1, \ldots, m$; $\underset{\sim}{\Pi}_i$ has $m+1$ rows and $m+1$ columns (including zeroth associates); the element in the $s^{th}$ row and $u^{th}$ column is $p_{si}^u$. The top (zero) row of $\underset{\sim}{\Pi}_i$ has unity in the $i^{th}$ column; its other entries are all zero. The left (zero) column has $n_i$ in the $i^{th}$ row and zero elsewhere; $\underset{\sim}{\Pi}_0$ is the identity matrix of order $m+1$. For the rectangular scheme we have:

$$\underset{\sim}{\mathrm{I\!I}}_1 = \begin{bmatrix} 0 & 1 & 0 & 0 \\ n-1 & n-2 & 0 & 0 \\ 0 & 0 & 0 & 1 \\ 0 & 0 & n-1 & n-2 \end{bmatrix}, \quad \underset{\sim}{\mathrm{I\!I}}_2 = \begin{bmatrix} 0 & 0 & 1 & 0 \\ 0 & 0 & 0 & 1 \\ m-1 & 0 & m-2 & 0 \\ 0 & m-1 & 0 & m-2 \end{bmatrix}.$$

The relationship just obtained from the triple products shows that the $(su)^{th}$ element of $\underset{\sim}{\mathrm{I\!I}}_i \underset{\sim}{\mathrm{I\!I}}_h$ is equal to the $(su)^{th}$ element of $\underset{\sim}{\mathrm{I\!I}}_h \underset{\sim}{\mathrm{I\!I}}_i$ which, in turn, is equal to the $(su)^{th}$ element of $\sum_k p_{hi\sim k}^k \underset{\sim}{\mathrm{I\!I}}_k$ so that

$$\underset{\sim}{\mathrm{I\!I}}_h \underset{\sim}{\mathrm{I\!I}}_i = \sum_k p_{hi\sim k}^k \underset{\sim}{\mathrm{I\!I}}_k = \underset{\sim}{\mathrm{I\!I}}_i \underset{\sim}{\mathrm{I\!I}}_h .$$

For the rectangular scheme

$$\underset{\sim}{\mathrm{I\!I}}_1 \underset{\sim}{\mathrm{I\!I}}_2 = \sum_k p_{12\sim k}^k \underset{\sim}{\mathrm{I\!I}}_k = \underset{\sim}{\mathrm{I\!I}}_3 = \underset{\sim}{\mathrm{I\!I}}_2 \underset{\sim}{\mathrm{I\!I}}_1 ,$$

$$\underset{\sim}{\mathrm{I\!I}}_1 \underset{\sim}{\mathrm{I\!I}}_3 = \sum_k p_{13\sim k}^k \underset{\sim}{\mathrm{I\!I}}_k = (n-1) \underset{\sim}{\mathrm{I\!I}}_2 + (n-2) \underset{\sim}{\mathrm{I\!I}}_3 = \underset{\sim}{\mathrm{I\!I}}_3 \underset{\sim}{\mathrm{I\!I}}_1 ,$$

$$\underset{\sim}{\mathrm{I\!I}}_2 \underset{\sim}{\mathrm{I\!I}}_3 = \sum_k p_{23\sim k}^k \underset{\sim}{\mathrm{I\!I}}_k = (m-1) \underset{\sim}{\mathrm{I\!I}}_1 + (m-2) \underset{\sim}{\mathrm{I\!I}}_3 = \underset{\sim}{\mathrm{I\!I}}_3 \underset{\sim}{\mathrm{I\!I}}_2 .$$

The matrices $\underset{\sim}{\mathrm{I\!I}}_i$ thus multiply in the same way as the association matrices; any polynomial equation satisfied by the $\underset{\sim}{B}_i$ is also satisfied by the $\underset{\sim}{\mathrm{I\!I}}_i$, and vice versa. In particular $\underset{\sim}{B}_i$ and $\underset{\sim}{\mathrm{I\!I}}_i$ have the same minimal polynomials so that, instead of having to find the $m+1$ distinct roots of a $t \times t$ matrix, we need only find the roots of a $(m+1) \times (m+1)$ matrix.

Since the association matrices commute, there exists an orthogonal matrix $\underset{\sim}{X}$ such that $\underset{\sim}{X}' \underset{\sim}{B}_i \underset{\sim}{X}$ is diagonal for each $i$. Each column of $\underset{\sim}{X}$ is a latent vector of each association matrix and, hence, of $\underset{\sim\sim}{NN'}$, $\underset{\sim}{C}$ and $\underset{\sim}{\Omega}$. One of the columns of $\underset{\sim}{X}$ is $\underset{\sim}{x}_0 = (t^{-1/2}) \underset{\sim}{1}$; let $\underset{\sim}{x}_u$ be any other column. We then have

$$\underset{\sim\sim}{NN'} \underset{\sim}{x}_u = r \underset{\sim}{I} \underset{\sim}{x}_u + \sum_{i=1}^m \lambda_i \underset{\sim}{B}_i \underset{\sim}{x}_u = \{r + \sum_{i=1}^m \lambda_i z_{ui}\} \underset{\sim}{x}_u = \theta_u \underset{\sim}{x}_u ;$$

$z_{ui}$ is the latent root of $\underset{\sim}{B}_i$ corresponding to $\underset{\sim}{x}_u$. Setting aside the simple root $\theta_0 = rk$, we may now write

$$\underset{\sim}{\theta} = r \underset{\sim}{1} + \underset{\sim}{Z}^* \underset{\sim}{\Lambda},$$

61

where $\underset{\sim}{\theta} = (\theta_1, \ldots, \theta_m)'$ is the vector of distinct latent roots of $\underset{\sim\sim}{NN'}$ and $\underset{\sim}{\Lambda} = (\lambda_1, \ldots, \lambda_m)'$. The elements of $\underset{\sim}{Z}^*$ are the latent roots of the association matrices; the $i^{th}$ column of $\underset{\sim}{Z}^*$ consists of the roots of $\underset{\sim}{\Pi}_i$ (other than a root $n_i$ which is included in $\theta_0$) in some order that we have yet to determine.

In the general situation $\underset{\sim\sim}{NN'}$ will have the full complement of $m+1$ distinct latent roots, and we proceed on that assumption. We shall see later that when certain of the $\lambda_i$ are equal some schemes degenerate. If, for example, in the rectangular scheme we have $\lambda_2 = \lambda_3$ we get a group divisible scheme.

9.5 Evaluating the roots of the concordance matrix. We have established that the coefficients of $\lambda_i$ in the various roots of $\underset{\sim\sim}{NN'}$ are the roots of $\underset{\sim}{\Pi}_i$ in some order. There is still a practical difficulty in finding which of the roots of $\underset{\sim}{\Pi}_i$ goes with $\underset{\sim}{\theta}_u$: which root is $z_{ui}$? In this section we shall show how $\underset{\sim}{\Pi}$ can be replaced by a matrix $\underset{\sim}{P}_i^*$ of order $m$ rather than $m+1$. We shall also present two devices that are often useful in reducing the amount of labor involved.

Consider the determinant $|\underset{\sim}{\Pi}_i - z\underset{\sim}{I}|$, where $z$ is a scalar. When all the other rows of the determinant are added to the top (zero) row, every element of that row becomes $n_i - z$. When the factor $n_i - z$ is extracted the top row becomes $\underset{\sim}{1}'$. The only other nonzero element in the zero column is $n_i$ in the $i^{th}$ row. We subtract $n_i\underset{\sim}{1}'$ from the $i^{th}$ row and we can now write

$$|\underset{\sim}{\Pi}_i - z\underset{\sim}{I}| = (n_i - z)\,|\underset{\sim}{P}_i^* - z\underset{\sim}{I}|$$

where $\underset{\sim}{P}_i^*$ is a matrix of $m$ rows and columns. The element in the $s^{th}$ row and the $u^{th}$ column of $\underset{\sim}{P}_i^*$ is $p^u_{si} - \delta_{si}n_i$. The remaining roots of $\underset{\sim}{B}_i$ are the roots of $\underset{\sim}{P}_i^*$.

Bose and Mesner (1959) note that the latent roots of $\underset{\sim\sim}{NN'}$ other than $rk$ are the roots of $\underset{\sim}{\Pi}^* = r\underset{\sim}{I} + \sum_{i=1}^{m} \lambda_i \underset{\sim}{P}_i^*$. The evaluation of the roots of $\underset{\sim}{\Pi}^*$ can, however, be tedious.

Lemma 9.5.1. $\underset{\sim}{Z}^*\underset{\sim}{1} = -\underset{\sim}{1}$.

Proof: Let $\underset{\sim}{x}_u$ be a column of $\underset{\approx}{X}$ with $\underset{\sim}{x}'\underset{\sim}{1} = 0$ corresponding to the roots $z_{ui}$. Then $\sum_i z_{ui}$ is a root of $\sum_{i=1}^{m} \underset{\sim}{B}_i$. However $\sum_{i=1}^{m} \underset{\sim}{B}_i = \underset{\sim}{J} - \underset{\sim}{I}$, and so $\sum z_{ui} = -1$ for each u.

For the rectangular association scheme we have:

$$\underset{\sim}{P}_1^* = \begin{bmatrix} -1 & -(n-1) & -(n-1) \\ 0 & 0 & 1 \\ 0 & (n-1) & (n-2) \end{bmatrix}, \quad \underset{\sim}{P}_2^* = \begin{bmatrix} 0 & 0 & 1 \\ -(m-1) & -1 & -(m-1) \\ (m-1) & 0 & (m-2) \end{bmatrix},$$

$$\underset{\sim}{P}_3^* = \begin{bmatrix} 0 & n-1 & n-2 \\ m-1 & 0 & m-2 \\ -(m-1) & -(n-1) & -m-n+3 \end{bmatrix}.$$

The latent roots of these matrices are: $\underset{\sim}{P}_1^*$ $-1, -1, n-1$; $\underset{\sim}{P}_2^*$ $-1, -1, m-1$; $\underset{\sim}{P}_3^*$ $-(m-1), -(n-1), 1$. Let us take the first of the latent roots of $\underset{\sim}{P}_3^*$ to go with the first root of $\underset{\sim}{N}\underset{\sim}{N}'$, i.e., let $z_{13} = -(m-1)$. In applying the lemma we must choose one of the latent roots of $\underset{\sim}{P}_1^*$ as $z_{11}$, and one of the latent roots of $\underset{\sim}{P}_2^*$ as $z_{12}$ in such a way that $z_{11} + z_{12} + z_{13} = -1$. This can only be done if $z_{11} = -1$ and $z_{12} = m-1$. It follows that the three roots other than rk are

$$\theta_1 = r - \lambda_1 + (m-1)\lambda_2 - (m-1)\lambda_3,$$

$$\theta_2 = r + (n-1)\lambda_1 - \lambda_2 - (n-1)\lambda_3,$$

$$\theta_3 = r - \lambda_1 + \lambda_2 - \lambda_3.$$

Lemma 9.5.2. The matrices $\underset{\sim}{P}_i^*$ commute.

The proof of this lemma involves calculating the elements in the $s^{th}$ row and $u^{th}$ column of $\underset{\sim}{P}_h^*\underset{\sim}{P}_i^*$ and $\underset{\sim}{P}_i^*\underset{\sim}{P}_h^*$ and observing that they are equal. The details are found in John (1978a) and will not be given here.

It follows from this lemma that if $z_{ui}$ is a simple root of $\underset{\sim}{P}_i^*$ with corresponding vector $\underset{\sim}{x}_u$, then $\underset{\sim}{x}_u$ is also a latent vector of $\underset{\sim}{P}_j^*$. We note that $\underset{\sim}{P}_i^*\underset{\sim}{P}_j^*\underset{\sim}{x}_u = \underset{\sim}{P}_j^* z_{ui}\underset{\sim}{x}_u = z_{ui}\underset{\sim}{P}_j^*\underset{\sim}{x}_u$, whence $\underset{\sim}{P}_j^*\underset{\sim}{x}_u$ is also a latent vector of $\underset{\sim}{P}_i^*$, and must be a scalar multiple of $\underset{\sim}{x}_u$. This result can be used whenever one of the $\tilde{P}_i^*$ has a simple latent root. In the example of the rectangular scheme, we could find the vectors corresponding to the three simple roots of $\tilde{P}_3^*$ and apply each of them in turn to $\tilde{P}_1^*$ and $\tilde{P}_2^*$.

9.6 <u>The multiplicities of the roots.</u> Let $\tilde{\alpha} = (\alpha_1, \ldots, \alpha_m)'$, where $\alpha_u$ is the multiplicity of $\theta_u$. Then the trace of $\tilde{N}\tilde{N}'$ is

$$\mathrm{tr}(\tilde{N}\tilde{N}') = rk + \tilde{\alpha}'\tilde{\theta} = rt,$$

so that

$$r(t - k) = \tilde{\alpha}'\tilde{Z}^*\tilde{\Lambda} + r\tilde{\alpha}'\tilde{1},$$

and, since $\tilde{\alpha}'\tilde{1} = t - 1$,

$$-r(k - 1) = \tilde{\alpha}'\tilde{Z}^*\tilde{\Lambda} = -\tilde{n}'\tilde{\Lambda}.$$

This is true for all designs with the scheme and is thus an identity in the $\lambda_i$. It follows that

$$\tilde{\alpha} = -(\tilde{Z}^{*\prime})^{-1}\tilde{n}.$$

For the rectangular scheme

$$\tilde{\alpha} = -(mn)^{-1}\begin{bmatrix} -n & 0 & -n \\ 0 & -m & -m \\ -n(m-1) & -m(n-1) & -mn+m+n \end{bmatrix}\begin{bmatrix} n-1 \\ m-1 \\ (m-1)(n-1) \end{bmatrix}$$

$$= (n - 1,\ m - 1,\ (m - 1)(n - 1))' = \tilde{n}.$$

We have already remarked that if $\lambda_2 = \lambda_3$ the scheme degenerates to the group divisible scheme. We have $\theta_1 = \theta_3 = r - \lambda_1$ and $\alpha_1 + \alpha_3 = m(n - 1)$. If $\lambda_1 = \lambda_2$ we have $\theta_1 = \theta_2$ only if $m = n$, in which case the rectangular scheme degenerates to the $L_2$ scheme; if $m \neq n$ we still have three distinct roots $\theta_1$, $\theta_2$, $\theta_3$ and three associate classes.

9.7 Efficiencies and intrablock estimates. The extension of the method that we used for the case of two associate classes is now straightforward. Since $\underset{\sim}{C}$ is a linear combination of the $\underset{\sim}{B}_i$, so is $\underset{\sim}{\Omega} = (\underset{\sim}{C} + a\underset{\sim}{J})^{-1}$; we write $\underset{\sim}{\Omega} = \omega_0\underset{\sim}{B}_0 + \sum_{i=1}^{m} \omega_i\underset{\sim}{B}_i$. We let $\underset{\sim}{\psi}$ be the vector whose $u^{th}$ element is $\psi_u = (rk - \theta_u)/k$, and write $(\underset{\sim}{\psi})^{-1}$ for the vector $(\psi_u^{-1})$. Let $\underset{\sim}{\omega} = (\omega_1, \ldots, \omega_m)'$. Then

$$(\underset{\sim}{\psi})^{-1} = \omega_0\underset{\sim}{1} + \underset{\sim}{Z}^*\underset{\sim}{\omega} = -\underset{\sim}{Z}^*(\omega_0\underset{\sim}{1} - \underset{\sim}{\omega}),$$

so that $\omega_0 - \omega_i$ is the $i^{th}$ element of $-\underset{\sim}{Z}^{*-1}(\underset{\sim}{\psi})^{-1}$. The efficiency of the estimate of the difference between two treatment effects is $E_i = r(\omega_0 - \omega_i)^{-1}$, if the treatments are $i^{th}$ associates. For the rectangular ES(3) design given earlier, $E_1 = 0.905$, $E_2 = 0.943$, $E_3 = 0.880$.

We can also extend the result for the intrablock analysis and obtain

$$\hat{\tau}_h = \omega_0 Q_h + \sum_{i=1}^{m} \omega_i S_i(Q_h) = -\sum(\omega_0 - \omega_i)S_i(Q_h).$$

# Chapter Ten

## SOME ASSOCIATION SCHEMES WITH MORE THAN TWO ASSOCIATE CLASSES

10.1 Introduction. In this chapter we shall present some association schemes with three or more associate classes, and give some examples of designs with these schemes. We begin with two schemes for $m = 3$, the hierarchic group divisible and the cubic. We then turn to the cyclic designs of J. A. John, Wolock and David (1972), and finally to the generalized cyclic designs and the Kronecker product designs, which come from the extended group divisible association scheme of Hinkelmann (1964).

10.2 The hierarchical group divisible scheme. This scheme was introduced by P. M. Roy (1953-1954), and considered more fully by Raghavarao (1960). We consider only the case in which $m = 3$. Let $t = N_1N_2N_3$. We divide the treatments into $N_1$ groups of $N_2N_3$ treatments each, and then divide each group into $N_2$ subgroups of $N_3$ treatments. Two treatments are said to be first associates if they are in the same subgroup, second associates if they are the same group, but in different subgroups, and third associates otherwise. This allocation is equivalent to taking a group divisible scheme and replacing each treatment by a subgroup of new treatments.

We have $n_1 = N_3 - 1$, $n_2 = N_3(N_2 - 1)$ and $n_3 = N_2N_3(N_1 - 1)$. The $\underset{\sim}{P}$ matrices are

$$\underset{\sim}{P}_1 = \begin{bmatrix} N_3 - 2 & 0 & 0 \\ 0 & n_2 & 0 \\ 0 & 0 & n_3 \end{bmatrix}, \quad \underset{\sim}{P}_2 = \begin{bmatrix} 0 & n_1 & 0 \\ n_1 & N_3(N_2 - 2) & 0 \\ 0 & 0 & n_3 \end{bmatrix},$$

$$\underset{\sim}{P}_3 = \begin{bmatrix} 0 & 0 & n_1 \\ 0 & 0 & n_2 \\ n_1 & n_2 & (N_1 - 2)N_2N_3 \end{bmatrix}.$$

The latent roots of $\underset{\sim\sim}{NN'}$ and their multiplicities are

$$\theta_1 = r - \lambda_3 + (\lambda_1 - \lambda_3)n_1 + (\lambda_2 - \lambda_3)n_2,$$

$$\theta_2 = r - \lambda_2 + (\lambda_1 - \lambda_2)n_1, \quad \theta_3 = r - \lambda_1, \quad \text{whence}$$

$$-\underset{\sim}{Z}^{*-1} = \Delta^{-1} \begin{bmatrix} 0 & 0 & (n_1 + 1)(n_1 + n_2 + 1) \\ 0 & n_1 + n_2 + 1 & n_1(n_1 + n_2 + 1) \\ n_1 + 1 & n_2 & n_1(n_1 + n_2 + 1) \end{bmatrix},$$

where $\Delta = (n_1 + 1)(n_1 + n_2 + 1)$, so that $\alpha_1 = N_1 - 1$, $\alpha_2 = N_1(N_2 - 1)$ and $\alpha_3 = $ $= N_1N_2(N_3 - 1)$.

There are two cases in which the scheme degenerates into a two-class scheme. If $\lambda_1 = \lambda_2$, then $\theta_2 = \theta_3$, and we have a group divisible scheme with $N_1$ groups of size $N_2N_3$. If, on the other hand, $\lambda_2 = \lambda_3$, then $\theta_1 = \theta_2$, and we have a group divisible scheme with $N_1N_2$ groups of $N_3$ treatments each. If, however, $\lambda_1 = \lambda_3$, we still have three distinct latent roots and the scheme does not degenerate.

We shall postpone our examples of designs with this scheme until later. It is clear that we can represent the treatments by three coordinates $(x_1, x_2, x_3)$, where $x_i = 0, 1, 2, \ldots, N_i - 1$. Two treatments are then first associates if they agree in both $x_1$ and $x_2$; they are second associates if they agree in $x_1$ but differ in $x_2$, and are third associates if they differ in $x_1$.

10.3 <u>The cubic association scheme</u>. This association scheme was introduced by Raghavarao and Chandrasekhararao (1964). There are $t = s^3$ treatments, each denoted by a triplet $(x_1, x_2, x_3)$, where $x_i = 0, 1, \ldots, s - 1$. Two treatments are first associates if they agree in exactly two coordinates, second associates if they have only one coordinate in common and third associates if all three coordinates differ. The $\underset{\sim}{P}$ matrices are

$$\underset{\sim}{P}_1 = \begin{bmatrix} s - 2 & 2(s - 1) & 0 \\ 2(s - 1) & 2(s - 1)(s - 2) & (s - 1)^2 \\ 0 & (s - 1)^2 & (s - 1)^2(s - 2) \end{bmatrix},$$

$$P_2 = \begin{bmatrix} 2 & 2(s-2) & s-1 \\ 2(s-2) & 2(s-1)+(s-2)^2 & 2(s-1)(s-2) \\ s-1 & 2(s-1)(s-2) & (s-1)(s-2)^2 \end{bmatrix},$$

$$P_3 = \begin{bmatrix} 0 & 3 & 3(s-2) \\ 3 & 6(s-2) & 3(s-2)^2 \\ 3(s-2) & 3(s-2)^2 & (s-2)^3 \end{bmatrix},$$

with $n_1 = 3(s-1)$, $n_2 = 3(s-1)^2$, $n_3 = (s-1)^3$.

The latent roots of $NN'$, in addition to the simple root $\theta_0 = rk$, are

$$\theta_1 = r + (2s-3)\lambda_1 + (s-1)(s-3)\lambda_2 - (s-1)^2\lambda_3,$$

$$\theta_2 = r + (s-3)\lambda_1 - (2s-3)\lambda_2 + (s-1)\lambda_3,$$

$$\theta_3 = r - 3\lambda_1 + 3\lambda_2 - \lambda_3.$$

We can now compute

$$-Z^{*-1} = s^{-2} \begin{bmatrix} 1 & 2(s-1) & (s-1)^2 \\ 2 & 3s-4 & (s-1)(s-2) \\ 3 & 3(s-2) & s^2-3s+3 \end{bmatrix}.$$

The multiplicities of the roots are thus $\alpha_1 = n_1$, $\alpha_2 = n_2$, $\alpha_3 = n_3$. The efficiencies are given by $(E_1^{-1}, E_2^{-1}, E_3^{-1})' = -Z^{*-1}(\psi)^{-1}r$.

For $t = 8$ the cubic scheme is the same as the rectangular scheme. We write the treatments in the array

| | | | |
|---|---|---|---|
| 000 | 011 | 101 | 110 |
| 111 | 100 | 010 | 001. |

The schemes are identical except that names of the associate classes are changed.

The following resolvable design, $D_1$, with $r = 3$, $k = 4$ has $\lambda_1 = 2$, $\lambda_2 = 1$, $\lambda_3 = 0$:

000, 100, 010, 110; 111, 011, 101, 001;

000, 100, 001, 101; 111, 011, 110, 010;

000, 010, 001, 011; 111, 101, 110, 100.

Another resolvable design, $D_2$, also has $r = 3$, $k = 4$; however, $\lambda_1 = \lambda_2 = 1$, $\lambda_3 = 3$. It is actually a singular group divisible design:

000, 111, 001, 110; 010, 101, 100, 011;

000, 111, 010, 101; 001, 110, 100, 011;

000, 111, 100, 011; 010, 101, 001, 110.

The single replicate,

000, 011, 101, 110; 111, 100, 010, 001,

forms a design, $D_3$, with $\lambda_1 = \lambda_3 = 0$, $\lambda_2 = 1$.

In the case where $s = 3$ and $t = 27$, the four initial blocks 000, 111, 222; 000, 112, 221; 000, 121, 212 and 000, 211, 112 give thirty-six blocks each repeated three times. Eliminating the duplicates gives a design, $D_4$, with $b = 36$, $r = 4$, $k = 3$, $\lambda_3 = 1$, $\lambda_1 = \lambda_2 = 0$. Each block consists of a set of treatments that are mutual third associates; there are exactly thirty-six such sets. The initial block 000, 011, 101, 110 gives a design, $D_5$, with $t = b = 27$, $r = k = 4$, $\lambda_2 = 1$, $\lambda_1 = \lambda_3 = 0$. The initial block 000, 111, 022, 202, 220 gives $D_6$ with $t = b = 27$, $r = k = 5$, $\lambda_1 = 0$, $\lambda_2 = \lambda_3 = 1$. The E'S(1) design, $D_7$, which is the symmetric design in which the $j^{th}$ block consists of the $j^{th}$ treatment together with its first associates, has $r = k = 7$, $\lambda_1 = 3$, $\lambda_2 = 2$, $\lambda_3 = 0$. The ES(3) design, $D_8$, in which the $j^{th}$ block consists of the third associates of the $j^{th}$ treatment, has $r = k = 8$, $\lambda_1 = 4$, $\lambda_2 = 2$, $\lambda_3 = 1$.

10.4 Cyclic designs. J. A. John, Wolock and David (1972) published a list of cyclic designs which were partially balanced designs with several associate classes. Their publication contains a bibliography which lists the earlier papers in the field by the three authors. The first use of these designs, with $k = 2$, was made by Kempthorne (1953) for breeding experiments.

The treatments are represented by the integers $0, 1, 2, \ldots, t-1$ with arithmetic $\bmod t$. The designs are obtained by developing cyclically one or more initial blocks. We have already used this method to obtain both balanced and group divisible designs. In the general case, however, the design will be partially balanced with more than two classes.

There are two cases to be considered. It $t$ is odd, we write $t = 2m + 1$. There are $m + 1$ associate classes; the $i^{th}$ associates of $0$ are $i$ and $t - i$; the $i^{th}$ associates of $h$ are $h + i$ and $h + t - i$. If $t$ is even, $t = 2m$, there are again $m$ classes. The first $m - 1$ classes are defined in the same way as when $t$ is odd, but the $m^{th}$ class now contains only one member. The $m^{th}$ associate of $h$ is $h + t/2$.

The concordance matrix, $\underset{\sim\sim}{NN'}$ is a circulant matrix. That property enables us to find the latent roots quickly. Let $\underset{\sim}{P}$ be a circulant matrix of order $t$ with the elements of the first row being $a_0, a_1, \ldots, a_{t-1}$. The $j^{th}$ latent root of $\underset{\sim}{P}$ is

$$\theta_j = a_0 + a_1\omega^j + a_2\omega^{2j} + \cdots + a_{t-1}\omega^{(t-1)j} \quad \text{for} \quad j = 0, 1, \ldots, t-1,$$

with corresponding vector $(1, \omega^j, \omega^{2j}, \ldots)'$, where $\omega^j = \cos(2\pi j/t) + \sqrt{(-1)}\sin(2\pi j/t)$; $\omega^j$ is a root of $x^t = 1$.

In the case of $\underset{\sim\sim}{NN'}$ we have $a_0 = r$ and $a_i = a_{t-i} = \lambda_i$. The imaginary parts of the latent roots vanish, since $\omega^{ij} + \omega^{t-ij} = 2\cos(2\pi ij/t)$, and we have, for $j = 1, 2, \ldots, m$,

$$\theta_j = r + \sum_{i=1}^{m} n_i\lambda_i\cos(2\pi ij/t).$$

The coefficients of $\lambda_i$ are real, but they may not be rational, as the following example shows.

<u>Example 10.4.1</u> $t = 8$.

The treatments are $0, 1, 2, 3, 4, 5, 6, 7$. The latent roots of $\underset{\sim\sim}{NN'}$ are $\theta_0 = rk$ with $\alpha_0 = 1$ and

$$\begin{aligned}
\theta_1 &= r + \sqrt{2}\lambda_1 \phantom{{}-2\lambda_2} - \sqrt{2}\lambda_3 - \lambda_4, & \alpha_1 &= 2,\\
\theta_2 &= r \phantom{{}+\sqrt{2}\lambda_1} - 2\lambda_2 \phantom{{}-\sqrt{2}\lambda_3} + \lambda_4, & \alpha_2 &= 1,\\
\theta_3 &= r - \sqrt{2}\lambda_1 \phantom{{}-2\lambda_2} + \sqrt{2}\lambda_3 - \lambda_4, & \alpha_3 &= 2,\\
\theta_4 &= r - 2\lambda_1 + 2\lambda_2 - 2\lambda_3 + \lambda_4, & \alpha_4 &= 2.
\end{aligned}$$

The $\underset{\sim}{P}$ matrices are

$$\underset{\sim}{P}_1 = \begin{bmatrix} 0 & 1 & 0 & 0 \\ 1 & 0 & 1 & 0 \\ 0 & 1 & 0 & 1 \\ 0 & 0 & 1 & 0 \end{bmatrix}, \quad \underset{\sim}{P}_2 = \begin{bmatrix} 1 & 0 & 1 & 0 \\ 0 & 0 & 0 & 1 \\ 1 & 0 & 1 & 0 \\ 0 & 1 & 0 & 0 \end{bmatrix},$$

$$\underset{\sim}{P}_3 = \begin{bmatrix} 0 & 1 & 0 & 1 \\ 1 & 0 & 1 & 0 \\ 0 & 1 & 0 & 0 \\ 1 & 0 & 0 & 0 \end{bmatrix}, \quad \underset{\sim}{P}_4 = \begin{bmatrix} 0 & 0 & 2 & 0 \\ 0 & 2 & 0 & 0 \\ 2 & 0 & 0 & 0 \\ 0 & 0 & 0 & 0 \end{bmatrix}.$$

It will be seen that when $\lambda_1 = \lambda_2 = \lambda_3$, we have $\theta_1 = \theta_2 = \theta_3$, and the design is group divisible. If $\lambda_1 = \lambda_3 \neq \lambda_2$, then $\theta_1 = \theta_3$, and the design is hierarchic group divisible.

10.5 Cyclic designs with the hierarchic GD scheme for $t = 8$. We illustrate this procedure for obtaining designs by some examples with the hierarchic group divisible scheme, $t = 8$, $N_1 = N_2 = N_3 = 2$. We take the subgroups of treatments to be 0, 4 and 2, 6 in the first group, and 1, 5 and 3, 7 in the second group.

The design, $D_9$, obtained by developing cyclically the initial block 0 1 4 has $b = 8$, $r = k = 3$ and $\lambda_1 = 2$, $\lambda_2 = 0$, $\lambda_3 = 1$. The initial block 0 1 5 gives a design with the same parameters. For these designs $\theta_1 = 1$, $\theta_2 = 5$, $\theta_3 = 1$ with $\alpha_1 = 1$, $\alpha_2 = 2$, $\alpha_3 = 4$. Although $\theta_1 = \theta_3$, there are three different values for $E_i$, namely $E_1 = 0.889$, $E_2 = 0.593$, $E_3 = 0.711$.

Taking 0 1 3 as the initial block gives design, $D_{10}$, with $\lambda_1 = 0$, $\lambda_2 = \lambda_3 = 1$; the initial block 0 1 6 gives a design with the same parameters. For these designs $\theta_1 = \theta_2 = 1$, $\theta_3 = 3$, $E_1 = 0.667$, $E_2 = E_3 = 0.762$. They are group divisible designs.

The initial block 0 2 4 gives another symmetric design, $D_{11}$, with $\lambda_1 = \lambda_2 = 2$, $\lambda_3 = 0$. This design is disconnected. Each block consists either of three treatments from the 'odd' group or three from the 'even' group.

We include now two designs with this hierarchic scheme that are not cyclic in the strict sense of our definition. They do, however, involve half cycles. We start with two initial blocks 0 1 7 and 0 2 5. We obtain six more blocks by adding in turn 2, 4 and 6 to the elements of each initial block. These new blocks together with the initial blocks give us a design with eight blocks of three plots each:

0 1 7, 2 3 1, 4 5 3, 6 7 5, 0 2 5, 2 4 7, 4 6 1, 6 0 3.

The blocks obtained by adding 1, 3, 5 and 7 in turn to the initial blocks form another design:

1 2 0, 3 4 2, 5 6 4, 7 0 6, 1 3 6, 3 5 0, 5 7 2, 7 1 4.

Both designs have the same parameters as $D_{10}$.

The seven designs that we have mentioned are disjoint, and they contain between them all fifty-six distinct blocks with three out of eight treatments. The smallest balanced incomplete block design for $t = 8$, $k = 3$ contains fifty-six blocks. If we take any one of the four designs with the parameters of $D_{10}$ and repeat it four times together with either of the $D_9$ designs twice, and add $D_{11}$ we have a balanced incomplete block design with $t = 8$, $k = 3$, $b = 56$, $\lambda = 6$, and only twenty-four distinct blocks. This design has been discussed by Wynn (1977) and John (1978b). The general topic of balanced incomplete block designs with repeated blocks has been investigated by Foody and Hedayat (1977); they call the subset of distinct blocks in a design the support of the design, and denote its size by $b^*$. Here we have $b = 56$, $b^* = 24$. Some other examples of balanced incomplete block designs with $t = 8$, $k = 3$, $b = 56$, $b^* = 32$, derived by piecing together group divisible designs, are also given by John (1978b).

With $k = 4$ we may take the initial block 0 1 2 5 and develop it to obtain a symmetric design with $\lambda_1 = \lambda_3 = 2$, $\lambda_2 = 1$. For this design $\theta_1 = 0$, $\theta_2 = 2$, $\theta_3 = 4$, $E_1 = 0.875$, $E_2 = 0.808$, $E_3 = 0.866$. It too may be incorporated in a balanced design, which has $k = 4$, $b = 14$, $\lambda = 3$. We add to this design, $D_{12}$, another design, $D_{13}$, which has $b = 6$, $\lambda_1 = \lambda_3 = 1$, $\lambda_2 = 2$; its blocks are

0 2 1 3, 4 6 1 7, 0 6 5 7, 2 4 5 3, 0 2 4 6, 1 3 5 7.

This design, $D_{12} + D_{13}$, is not resolvable. If we go back for a moment to Section 10.3 and add designs $D_1$, $D_2$ and $D_3$ we obtain a balanced design with the same parameters, which is resolvable. These two balanced designs are thus not isomorphic. Furthermore, it will be recalled that the dual of a resolvable balanced incomplete blcok design is a semiregular group divisible design. This is the case with the dual of $D_1 + D_2 + D_3$. Although $D_{12} + D_{13}$ has the same parameters, its dual is not partially balanced.

10.6 <u>Generalized cyclic designs</u>. In the cyclic designs, the treatments are represented by the integers $\bmod t$, and $\underset{\sim}{N}\underset{\sim}{N}'$ is a circulant matrix. We have already seen that, if $t = mn$, we can represent the treatments by two coordinates $x, y$, where $x = 0, 1,\ldots,m-1$ and $y = 0, 1,\ldots,n-1$, and obtain designs by developing initial blocks. These designs and similar designs with three or more coordinates are called generalized cyclic designs. The discussion will be confined to a large class of designs and schemes in which $\underset{\sim}{N}\underset{\sim}{N}'$ again has a special structure. We shall start this section by re-examining the rectangular association scheme from a somewhat different point of view. In subsequent sections, we shall consider representing treatments by three coordinates, and we shall obtain the hierarchic group divisible and cubic schemes as special cases.

Let $t = N_1N_2$. The treatments are denoted by two coordinates $x_1$, $x_2$, $x_1 = 0, 1,\ldots,N_1 - 1$ and $x_2 = 0, 1,\ldots,N_2 - 1$. There are three associate classes. Two treatments are 01 associates if their representations have the same value for $x_1$, but different values for $x_2$; they are 10 associates if they have the same $x_2$, but different $x_1$, and 11 associates if both coordinates differ. This is no more than the rectangular scheme with $N_1$ for m, and $N_2$ for n. We have $n_{01} = N_2 - 1$, $n_{10} = N_1 - 1$, $n_{11} = (N_1-1)(N_2 - 1)$.

If the treatments are ordered lexicographically, 00, 01,...,10, 11, ...,$(N_1 - 1)(N_2 - 1)$, the concordance matrix has a special structure. If we partition $\underset{\sim}{N}\underset{\sim}{N}'$ into $N_1^2$ square submatrices of order $N_2$ we see that the submatrices along the main diagonal are $\underset{\sim}{A}_0 = (r - \lambda_{01})\underset{\sim}{I} + \lambda_{01}\underset{\sim}{J}$; the off-diagonal matrices are $\underset{\sim}{A}_1 = (\lambda_{10} - \lambda_{11})\underset{\sim}{I} + \lambda_{11}\underset{\sim}{J}$.

That is how Hinkelmann (1964) saw the association scheme. Vartak (1955) arrived there by a different route. He considered the situation in which $\underset{\sim}{N}\underset{\sim}{N}'$ is the Kronecker product of the concordance matrices of two balanced incomplete block designs, which is a special case of the rectangular scheme.

The reader may recall that if $\underset{\sim}{A} = (a_{ij})$ and $\underset{\sim}{B}$ are two matrices of order $N_1$ and $N_2$ respectively the Kronecker, or direct, product $\underset{\sim}{A} \times \underset{\sim}{B}$ is a matrix of order $N_1N_2$:

$$\underset{\sim}{A} \times \underset{\sim}{B} = \begin{pmatrix} a_{11}\underset{\sim}{B} & a_{12}\underset{\sim}{B} & \cdots\cdots \\ a_{21}\underset{\sim}{B} & a_{22}\underset{\sim}{B} & \cdots\cdots \end{pmatrix}$$

For the rectangular scheme, as Vartak saw it in his paper, $\underset{\sim\sim}{NN'}$ has the structure

$$\underset{\sim\sim}{NN'} = ((a-b)\underset{\sim}{I}_1 + b\underset{\sim}{J}_1) \times ((c-d)\underset{\sim}{I}_2 + d\underset{\sim}{J}_2),$$

where $\underset{\sim}{I}_i$, $\underset{\sim}{J}_i$ denotes a matrix of order $N_i$; we shall see later that this is not a necessary condition on $\underset{\sim\sim}{NN'}$ for designs with the rectangular scheme. The latent roots of $\underset{\sim\sim}{NN'}$, which we obtained in chapter nine,may now be rewritten as $\theta_{00} = rk$ with $\alpha_{00} = 1$ and

$$\begin{aligned} \theta_{10} &= r - \lambda_{10} + n_{01}(\lambda_{01} - \lambda_{11}), & \alpha_{10} &= N_1 - 1, \\ \theta_{01} &= r - \lambda_{01} + n_{10}(\lambda_{10} - \lambda_{11}), & \alpha_{01} &= N_2 - 1, \\ \theta_{11} &= r - \lambda_{01} - \lambda_{10} + \lambda_{11}, & \alpha_{11} &= (N_1 - 1)(N_2 - 1). \end{aligned}$$

When $\lambda_{01} = \lambda_{11}$, or when $\lambda_{10} = \lambda_{11}$, the scheme degenerates to the group divisible scheme. When $\lambda_{01} = \lambda_{10}$ and $N_1 = N_2$ we have the $L_2$ scheme.

10.7 <u>Hinkelmann's EGD schemes.</u> Hinkelmann (1964) extended this idea to several coordinates in his EGD, extended group divisible schemes. He had originally introduced them with Kempthorne (1963) in connection with breeding experiments. We shall consider the case of three coordinates. The reader who wishes to go further is referred to Hinkelmann's paper.

Let $t = N_1N_2N_3$. We can now represent each treatment by three coordinates $x_1, x_2, x_3$ with $x_i = 0, 1, \ldots, N_i - 1$. There are seven associate classes which we can denote by $(a_1, a_2, a_3)$; two treatments are $(a_1, a_2, a_3)$ associates with $a_i = 0$ if $x_i$ is the same for both treatments, and $a_i = 1$ if the values of $x_i$ differ.

If we now include a row and column for 000 associates in the $\underset{\sim}{P}$ matrices, they are obtained in the following way. Let

$$\underset{\sim}{P}_0(i) = \begin{bmatrix} 1 & 0 \\ 0 & N_i - 1 \end{bmatrix}, \qquad \underset{\sim}{P}_1(i) = \begin{bmatrix} 0 & 1 \\ 1 & N_i - 2 \end{bmatrix}.$$

Then $\mathbf{P}_{ijk} = \mathbf{P}_i(1) \times \mathbf{P}_j(2) \times \mathbf{P}_k(3)$. The reader may wish to check this for the rectangular scheme with two coordinates.

The class sizes are $n_{100} = N_1 - 1$, $n_{010} = N_2 - 1$, $n_{001} = N_3 - 1$, $n_{011} = (N_2 - 1)(N_3 - 1)$, $n_{101} = (N_1 - 1)(N_3 - 1)$, $n_{110} = (N_1 - 1)(N_2 - 1)$, $n_{111} = (N_1 - 1(N_2 - 1)(N_3 - 1)$.

We now write the treatments in the order 000, 001, 002,...,010, 011, ...,100, 101,...,$(N_1 - 1)(N_2 - 1)(N_3 - 1)$. The concordance matrix $\mathbf{NN}'$ then has a particular cyclic structure. We first partition it into $(N_1)^2$ square submatrices of order $N_2N_3$; these submatrices are $\mathbf{A}_0$ along the main diagonal and $\mathbf{A}_1$ off the diagonal. We then partition $\mathbf{A}_0$ into square matrices of order $N_3$; along the main diagonal we have $\mathbf{A}_{00} = (r - \lambda_{001})\mathbf{I} + \lambda_{001}\mathbf{J}$; off the diagonal we have $\mathbf{A}_{01} = (\lambda_{010} - \lambda_{011})\mathbf{I} + \lambda_{011}\mathbf{J}$. We also partition $\mathbf{A}_1$ into square submatrices of order $N_3$; along the main diagonal $\mathbf{A}_{10} = (\lambda_{100} - \lambda_{101})\mathbf{I} + \lambda_{101}\mathbf{J}$ and off the diagonal $\mathbf{A}_{11} = (\lambda_{110} - \lambda_{111})\mathbf{I} + \lambda_{111}\mathbf{J}$. There are seven distinct latent roots of $\mathbf{NN}'$ in addition to the simple root $\theta_{000} = rk$. They are

$$\theta_{001} = r - \lambda_{001} + n_{010}(\lambda_{010} - \lambda_{011}) + n_{100}(\lambda_{100} - \lambda_{101}) + n_{110}(\lambda_{110} - \lambda_{111}), \qquad \alpha_{001} = n_{001},$$

$$\theta_{010} = r - \lambda_{010} + n_{001}(\lambda_{001} - \lambda_{011}) + n_{100}(\lambda_{100} - \lambda_{110}) + n_{101}(\lambda_{101} - \lambda_{111}), \qquad \alpha_{010} = n_{010},$$

$$\theta_{100} = r - \lambda_{100} + n_{010}(\lambda_{010} - \lambda_{110}) + n_{001}(\lambda_{001} - \lambda_{101}) + n_{011}(\lambda_{011} - \lambda_{111}), \qquad \alpha_{100} = n_{100},$$

$$\theta_{110} = r - \lambda_{100} - \lambda_{010} + \lambda_{110} + n_{001}(\lambda_{001} - \lambda_{101} - \lambda_{011} + \lambda_{111}), \qquad \alpha_{110} = n_{110},$$

$$\theta_{101} = r - \lambda_{100} - \lambda_{001} + \lambda_{101} + n_{010}(\lambda_{010} - \lambda_{110} - \lambda_{011} + \lambda_{111}), \qquad \alpha_{101} = n_{101},$$

$$\theta_{011} = r - \lambda_{010} - \lambda_{001} + \lambda_{011} + n_{100}(\lambda_{100} - \lambda_{110} - \lambda_{101} + \lambda_{111}), \qquad \alpha_{011} = n_{011},$$

$$\theta_{111} = r - \lambda_{100} - \lambda_{010} - \lambda_{001} + \lambda_{110} + \lambda_{101} + \lambda_{011} - \lambda_{111}, \qquad \alpha_{111} = n_{111}.$$

When some of the $\lambda_{ijk}$ and some of the $N_i$ are equal, the EGD scheme degenerates. The hierarchical group divisible scheme is obtained when we let $\lambda_1 = \lambda_{001}$, $\lambda_2 = \lambda_{010} = \lambda_{011}$ and $\lambda_3 = \lambda_{100} = \lambda_{101} = \lambda_{110} = \lambda_{111}$. The cubic scheme is obtained when we let $N_1 = N_2 = N_3 = s$ and let $\lambda_1 = \lambda_{100} = \lambda_{010} = \lambda_{001}$, $\lambda_2 = \lambda_{011} = \lambda_{101} = \lambda_{110}$ and $\lambda_3 = \lambda_{111}$.

The extended rectangular scheme is obtained when we let $\lambda_1 = \lambda_{100}$, $\lambda_2 = \lambda_{010} = \lambda_{110}$, $\lambda_3 = \lambda_{001} = \lambda_{101}$ and $\lambda_4 = \lambda_{011} = \lambda_{111}$. If $N_2 = N_3$ and $\lambda_2 = \lambda_3$ we have the extended $L_2$ scheme of Singla (1977).

10.8 Kageyama's derivation of the latent roots. Kageyama (1972), following Vartak, considered designs for which $\underset{\sim}{N}\underset{\sim}{N}'$ is the Kronecker product of the concordance matrices of m balanced incomplete block designs. He calls the corresponding schemes $F_m$ association schemes; they are the same as Hinkelmann's EGD schemes. Kageyama's method of obtaining the latent roots of $\underset{\sim}{N}\underset{\sim}{N}'$ is interesting, and we present it for the case $m = 2$. The reader may then find it instructive to use his method to derive the roots for the case $m = 3$ that were given in the previous section.

We denote the parameters of the two BIB designs by $t_i$, $r_i$, $k_i$, $\mu_i$ (rather than $\lambda_i$) and their incidence matrices by $\underset{\sim}{N}_i$, $i = 1, 2$. We consider a design, D, whose incidence matrix is $\underset{\sim}{N} = \underset{\sim}{N}_1 \times \underset{\sim}{N}_2$. There are $t = t_1 t_2$ treatments in D, corresponding to pairs of original treatments, one from each BIBD. The design has $r = r_1 r_2$ and $k = k_1 k_2$.

We recall that $(\underset{\sim}{A} \times \underset{\sim}{B})' = \underset{\sim}{A}' \times \underset{\sim}{B}'$ and that $(\underset{\sim}{A} \times \underset{\sim}{B})(\underset{\sim}{C} \times \underset{\sim}{D}) = \underset{\sim}{A}\underset{\sim}{C} \times \underset{\sim}{B}\underset{\sim}{D}$. The concordance matrix of D is thus

$$\underset{\sim}{N}\underset{\sim}{N}' = \underset{\sim}{N}_1\underset{\sim}{N}_1' \times \underset{\sim}{N}_2\underset{\sim}{N}_2' = ((r_1 - \mu_1)\underset{\sim}{I} + \mu_1\underset{\sim}{J}) \times ((r_2 - \mu_2)\underset{\sim}{I} + \mu_2\underset{\sim}{J}).$$

This is the concordance matrix of a design with the rectangular scheme and $\lambda_{11} = \mu_1\mu_2$, $\lambda_{01} = r_1\mu_2$, $\lambda_{10} = \mu_1 r_2$ and $r = r_1 r_2$.

For each i there exists an orthogonal matrix $\underset{\sim}{H}_i$ such that $\underset{\sim}{H}_i\underset{\sim}{N}_i\underset{\sim}{N}_i'\underset{\sim}{H}_i' = \underset{\sim}{G}_i$, where $\underset{\sim}{G}_i$ is a diagonal matrix whose diagonal elements are the latent roots of $\underset{\sim}{N}_i\underset{\sim}{N}_i'$. The diagonal of $\underset{\sim}{G}_i$ has $r_i k_i$ in the first position; the other diagonal elements are all equal to $r_i - \mu_i$. The matrix

$$(\underset{\sim}{H}_1 \times \underset{\sim}{H}_2)\underset{\sim}{N}\underset{\sim}{N}'(\underset{\sim}{H}_1 \times \underset{\sim}{H}_2)' = \underset{\sim}{G} = \underset{\sim}{G}_1 \times \underset{\sim}{G}_2$$

is also diagonal. Its diagonal elements are the latent roots of $\underset{\sim}{N}\underset{\sim}{N}'$ which are thus

$$\theta_{00} = r_1k_1r_2k_2 = rk, \qquad \alpha_{00} = 1,$$

$$\theta_{10} = (r_1 - \mu_1)r_2k_2, \qquad \alpha_{10} = t_1 - 1,$$

$$\theta_{01} = r_1k_1(r_2 - \mu_2), \qquad \alpha_{01} = t_2 - 1,$$

$$\theta_{11} = (r_1 - \mu_1)(r_2 - \mu_2), \qquad \alpha_{11} = (t_1 - 1)(t_2 - 1).$$

The expressions that we gave for $\theta_{10}$, $\theta_{01}$ and $\theta_{11}$ in section 10.7 are now derived by substituting $\lambda_{11}$ for $\mu_1\mu_2$ etc. We derive $\theta_{01}$ as an example: we note that from the properties of balanced incomplete block designs

$$r_1k_1 = r_1(k_1 - 1) + r_1 = r_1 + (t_1 - 1)\mu_1.$$

Then

$$\theta_{01} = r_1k_1(r_2 - \mu_2) = r_1r_2 - r_1\mu_2 + (t_1 - 1)(r_2\mu_1 - \mu_1\mu_2)$$

$$= r - \lambda_{01} + n_{10}(\lambda_{10} - \lambda_{11}).$$

It is not true that every rectangular design must be the Kronecker product of two balanced designs. Consider the following counter-example. The complete lattice for $t = 9$, $k = 3$ has four replications

| | | | |
|---|---|---|---|
| 1 2 3 | 1 4 7 | 1 5 9 | 1 6 8 |
| 4 5 6 | 2 5 8 | 2 6 7 | 2 4 9 |
| 7 8 9 | 3 6 9 | 3 4 8 | 3 5 7. |

Take the first replication to indicate the rectangular array of the association scheme. Let D be the design with the last two replicates only: $r = 2$, $k = 3$, $\lambda_{01} = \lambda_{10} = 0$, $\lambda_{11} = 1$. If we have square matrices $(a - b)\underset{\sim}{I} + b\underset{\sim}{J}$ and $(c - d)\underset{\sim}{I} + d\underset{\sim}{J}$ such that their Kronecker product is equal to $\underset{\sim\sim}{NN'}$, then $ac = 2$, $ad = bc = 0$, $bd = 1$. This would imply $a/b = ac/bc = 2/0$, and $a/b = ad/bd = 0/1$. Indeed, a necessary condition for $\underset{\sim\sim}{NN'}$ to be the Kronecker product of two such matrices is that $r\lambda_{11} = \lambda_{10}\lambda_{01}$.

10.9 Combining associate classes. We have already seen that in some instances, when $\lambda_h = \lambda_i$, an association scheme will degenerate to a scheme with $(m-1)$ classes. Our approach has always been to note that in this case two of the latent roots of $\underset{\sim}{N}\underset{\sim}{N}'$ are identical. An alternative approach which uses only the $\underset{\sim}{P}$ matrices was given by Vartak (1955). The proof is obvious.

We consider for convenience the case in which $\lambda_1 = \lambda_2$. Let us obtain a new matrix $\underset{\sim}{Q}_i$ from $\underset{\sim}{P}_i$ in two stages; we add the second row to the first row and remove the second row; we then add the second column to the first column and remove the second column. The matrix $\underset{\sim}{Q}_i$ has $(m-1)$ rows and $(m-1)$ columns. Its elements are

$$q^i_{11} = p^i_{11} + p^i_{12} + p^i_{21} + p^i_{22}; \quad q^i_{j1} = p^i_{j+1,1} + p^i_{j+1,2};$$

$$q^i_{jk} = p^i_{j+1,k+1}, \quad \text{when } j > 2 \text{ and } k > 2.$$

A necessary and sufficient condition that pooling the first and second associate classes will produce an association scheme with $(m-1)$ classes is that $\underset{\sim}{Q}_1 = \underset{\sim}{Q}_2$. Applying this to the cyclic designs with $t = 8$, $m = 4$ that were discussed in Section 10.4, we see that, if $\lambda_1 = \lambda_3$,

$$\underset{\sim}{Q}_1 = \underset{\sim}{Q}_3 = \begin{bmatrix} 0 & 2 & 1 \\ 2 & 0 & 0 \\ 1 & 0 & 0 \end{bmatrix}, \quad \underset{\sim}{Q}_2 = \begin{bmatrix} 4 & 0 & 0 \\ 0 & 0 & 1 \\ 0 & 1 & 0 \end{bmatrix}, \quad \underset{\sim}{Q}_4 = \begin{bmatrix} 4 & 0 & 0 \\ 0 & 2 & 0 \\ 0 & 0 & 0 \end{bmatrix}.$$

On the other hand we recall that in design $D_{12}$ of Section 10.5, which had the hierarchic group divisible scheme for $t = 8$, there were three distinct latent roots for $\underset{\sim}{N}\underset{\sim}{N}'$ (and three efficiencies), even though $\lambda_1 = \lambda_3$. For that scheme

$$\underset{\sim}{P}_1 = \begin{bmatrix} 0 & 0 & 0 \\ 0 & 2 & 0 \\ 0 & 0 & 4 \end{bmatrix}, \quad \underset{\sim}{P}_2 = \begin{bmatrix} 0 & 1 & 0 \\ 1 & 0 & 0 \\ 0 & 0 & 4 \end{bmatrix}, \quad \underset{\sim}{P}_3 = \begin{bmatrix} 0 & 0 & 1 \\ 0 & 0 & 2 \\ 1 & 2 & 0 \end{bmatrix}.$$

If we try to pool the first and third classes we obtain

$$\underset{\sim}{Q}_1 = \begin{bmatrix} 4 & 0 \\ 0 & 2 \end{bmatrix}, \qquad \underset{\sim}{Q}_2 = \begin{bmatrix} 4 & 1 \\ 1 & 0 \end{bmatrix}, \qquad \underset{\sim}{Q}_3 = \begin{bmatrix} 2 & 2 \\ 2 & 0 \end{bmatrix},$$

and $\underset{\sim}{Q}_1 \neq \underset{\sim}{Q}_3$.

## Chapter Eleven

# MORE ABOUT VARIANCES AND EFFICIENCIES

11.1 Introduction. The overall intrablock efficiency of an incomplete block design is denoted by $E$. It is usually defined as the reciprocal of the ratio of the average variance of the simple comparisons $\hat{\tau}_h - \hat{\tau}_i$ from the design to the value that the average would have taken if we had the same number of plots in a randomized complete block design. The latter quantity is $2\sigma^2/\bar{r}$, where $n$ is the total number of plots and $\bar{r} = n/t$.

To evaluate the denominator of $E$, we write $\underset{\sim}{\Omega} = (\omega_{hi})$ and recall that

$$\sigma^{-2}V(\hat{\tau}_h - \hat{\tau}_i) = \omega_{hh} + \omega_{ii} - 2\omega_{hi}.$$

The average variance, $\bar{v}$, of a comparison is thus given by

$$\sigma^{-2}t(t-1)\bar{v}/2 = \sum_{h=1}^{t-1}\sum_{i=h+1}^{t}(\omega_{hh} + \omega_{ii} - \omega_{hi} - \omega_{ih})$$

$$= \sum_{h=1}^{t}(t\omega_{hh} - \sum_{i=1}^{t}\omega_{hi}) = t(\mathrm{tr}\underset{\sim}{\Omega}) - \underset{\sim}{1}'\underset{\sim}{\Omega}\underset{\sim}{1},$$

where $\mathrm{tr}\underset{\sim}{\Omega}$ is the trace of $\underset{\sim}{\Omega}$.

We now recall that $\underset{\sim}{\Omega} = (\underset{\sim}{C} + a\underset{\sim}{J})^{-1}$, where $a$ is some scalar, and that $\underset{\sim}{\Omega}\underset{\sim}{1} = (at)^{-1}\underset{\sim}{1}$. The other latent roots of $\underset{\sim}{\Omega}$ are $\psi_i^{-1}$, the reciprocals of the nonzero roots of $\underset{\sim}{C}$. It then follows directly that

$$t(t-1)\bar{v}/2 = t\sum(\psi_i^{-1})\sigma^2 \quad \text{and} \quad E = (t-1)/\{\bar{r}\sum(\psi_i^{-1})\}.$$

For an equireplicate design, $\bar{r} = r$. For a partially balanced design,

$$(t-1)E^{-1} = \sum n_i E_i^{-1}.$$

The quantity $\bar{r}E$ is the harmonic mean of the $\psi_i$. It takes its maximum value, subject to the constraint $\sum \psi_i = \mathrm{tr}\,\underset{\sim}{C}$, when all the $\psi_i$ are equal. Kempthorne (1956) conjectured that, for any proper equireplicate designs with given $t, b, r, k$, the most efficient design is a balanced incomplete block design if one exists. His conjecture was proved by J. Roy (1958). This gives an upper bound to the efficiency

$$E_0 = \{t(k-1)\}/\{k(t-1)\}.$$

11.2 Designs with $t > b$. Patterson and Williams (1976) obtained a smaller upper bound than $E_0$ for proper binary equireplicate designs with $t > b$. The matrices $\underset{\sim}{N}\underset{\sim}{N}'$ and $\underset{\sim}{N}'\underset{\sim}{N}$ have the same positive latent roots with the same multiplicities; the rank of each matrix is no greater than the lesser of $t$ and $b$. If, therefore, a design has $t = b + s$, we know that zero is a latent root of $\underset{\sim}{N}\underset{\sim}{N}'$ with multiplicity at least $s$. To each appearance of zero as a root of $\underset{\sim}{N}\underset{\sim}{N}'$ corresponds a root $\psi = r$ of $\underset{\sim}{C}$.

The maximum value of $\sum(\psi_i^{-1})$ is now obtained subject to this additional restriction. It occurs when $\psi_i = r$ with multiplicity $s$, and $\psi_i = \{rt(r-1)\}/\{k(b-1)\}$ with multiplicity $b-1$, so that

$$E = E_0' = \frac{t(t-1)(r-1)}{t(t-1)(r-1) + (b-1)(t-k)} .$$

If, in addition, the design is resolvable, the multiplicity of the root $\psi = r$ is increased by $r - 1$, and their result becomes

$$E = E_0'' = \frac{(t-1)(r-1)}{(t-1)(r-1) + (b-r)}$$

This result implies that the design is a linked block design because $\theta$ takes only three values: $\theta_0 = rk$ with $\alpha_0 = 1$, $\theta_1 = 0$ with $\alpha_1 = t - b$, and $\theta_2$ with $\alpha_2 = b - 1$. Thus its dual has concordance matrix $\underset{\sim}{N}'\underset{\sim}{N}$ with latent roots $\theta_0^* = rk$, once, and $\theta_1^* = \theta_2$, $(b-1)$ times; it is therefore a balanced incomplete block design. Hence, if $t > b$, the most efficient design will be a linked block design, if one exists. Linked block designs are defined in Section 7.2.

11.3 Latent vectors of the concordance matrix. We have already seen that for a proper equireplicate design the latent vectors of $\underset{\sim}{N}\underset{\sim}{N}'$ are also the latent vectors of $\underset{\sim}{\Omega}$. Let $\underset{\sim}{c}$ denote any latent vector of $\underset{\sim}{\Omega}$ other than $\underset{\sim}{1}$ with $\psi^{-1}$ as the corresponding root; then $\underset{\sim}{c}'\hat{\underset{\sim}{\tau}}$ is a contrast and $V(\underset{\sim}{c}'\hat{\underset{\sim}{\tau}}) = \psi^{-1}\underset{\sim}{c}'\underset{\sim}{c}\sigma^2$. If $\underset{\sim}{c}_h$ and $\underset{\sim}{c}_i$ are orthogonal latent vectors, which will, in particular, be the case if they correspond to different roots, $\text{cov}(\underset{\sim}{c}_h'\hat{\underset{\sim}{\tau}}, \underset{\sim}{c}_i'\hat{\underset{\sim}{\tau}}) = 0$.

We can compute an intrablock efficiency for the contrast $\underset{\sim}{c}'\hat{\underset{\sim}{\tau}}$ by comparing $\psi^{-1}\underset{\sim}{c}'\underset{\sim}{c}$ to $\underset{\sim}{c}'\underset{\sim}{c}/r$. This gives $E^* = \psi/r$. If we now substitute $\psi = r - (\theta/k)$, where $\theta$ is the root of $\underset{\sim}{N}\underset{\sim}{N}'$ corresponding to $\psi$, we obtain

$$E^* = 1 - (\theta/rk).$$

The quantity $\theta/rk$ can be taken as a measure of the loss of efficiency for the contrast in the design.

11.4 Other criteria for efficiency. Kiefer (1959) has suggested several criteria for choosing between designs. His primary concern has been with optimal designs, and his approach has been that of decision theory. Among his criteria, three are of more interest than the others: A design is said to be:

A-optimal if the quantity $(\sum\psi_i^{-1})$ is minimized.

D-optimal if $\Pi\psi_i$, the product of the nonzero roots of $\underset{\sim}{C}$, is maximized.

E-optimal if $(\min \psi_i)$ is maximized.

The first of these, which is sometimes called trace optimality, corresponds to the intrablock efficiency, E, that we have discussed in the previous sections. The second, which minimizes the determinant of $\underset{\sim}{\Omega}$, is akin to minimizing a generalized variance for all linear combinations of the $\tau_i$. The third involves minimizing the largest of the variances of the linear combinations $(\underset{\sim}{c}'\underset{\sim}{c})^{-\frac{1}{2}}\underset{\sim}{c}'\hat{\underset{\sim}{\tau}}$, where $\underset{\sim}{c}$ is a latent vector of $\underset{\sim}{N}\underset{\sim}{N}'$.

Jarrett (1977) investigates an alternative measure of efficiency based on the canonical efficiencies of James and Wilkinson (1971). These are the $t - 1$ nonzero latent roots of the matrix $\underset{\sim}{R}^{-\frac{1}{2}}\underset{\sim}{C}\underset{\sim}{R}^{-\frac{1}{2}}$, and are denoted by $e_i$. Jarrett and also Williams (1975) use an efficiency factor defined as the harmonic mean of the $e_i$:

$$\bar{E} = (t-1)/(\sum e_i^{-1}).$$

For equireplicate designs this gives the same values as the efficiency factor that we have used, because in this case $\underset{\sim}{R} = r\underset{\sim}{I}$ and $e_i = \psi_i/r$. For unequal replicates it gives different values.

Pearce (1968, 1970) has considered the mean efficiency of adjusted means, which he denotes by $\mathscr{E}$. We shall now derive his criterion for equireplicate designs. His concern is with the efficiency of the estimate $\hat{\mu} + \hat{\tau}_h = \bar{y} + \hat{\tau}_h$, where $\bar{y}$ is the grand mean of the data. We let $V_h$ denote the variance of $(\bar{y} + \hat{\tau}_h)$. Since $\mathrm{cov}(\bar{y}, Q_i) = 0$ for each $i$, $\mathrm{cov}(\bar{y}, \hat{\tau}_h) = 0$, and so $\sigma^{-2}V_h = (rt)^{-1} + V(\hat{\tau}_h)$. The covariance matrix of the estimates $\hat{\tau}_h$ is given by

$$\sigma^{-2}\mathrm{cov}(\hat{\underset{\sim}{\tau}}) = \sigma^{-2}\mathrm{cov}(\underset{\sim}{\Omega}\underset{\sim}{Q}) = \underset{\sim}{\Omega}\underset{\sim}{C}\underset{\sim}{\Omega}.$$

We saw earlier that $\underset{\sim}{C}\underset{\sim}{\Omega} = \underset{\sim}{I} - \underset{\sim}{J}/t$, and so

$$\underset{\sim}{\Omega}\underset{\sim}{C}\underset{\sim}{\Omega} = \underset{\sim}{\Omega} - \underset{\sim}{\Omega}\underset{\sim}{J}/t = \underset{\sim}{\Omega} - (at^2)^{-1}\underset{\sim}{J},$$

whence $V(\hat{\tau}_h) = \omega_{hh} - (at^2)^{-1}$;

$$\sum V_h = r^{-1} + \mathrm{tr}\underset{\sim}{\Omega} - (at)^{-1} = r^{-1} + \sum \psi^{-1}.$$

Then, since $\mathscr{E} = t/(r\sum V_h)$,

$$t/\mathscr{E} = 1 + r\sum\psi^{-1} = 1 + (t-1)/E.$$

It follows that

$$\mathscr{E} = \frac{Et}{E + t - 1}, \qquad E = \frac{\mathscr{E}(t-1)}{t - \mathscr{E}}.$$

We note that $\mathscr{E} \geq E$ with equality only if $E = 1$.

# Chapter Twelve

# FACTORIAL EXPERIMENTS IN INCOMPLETE BLOCKS

12.1 Introduction. The basic ideas of the factorial experiment and confounding are given in most standard books. For example, we may wish to carry out a series of experimental runs on a pilot plant varying both temperature and pressure. Suppose that we elect to run the plant at $a$ different temperatures and at $b$ different pressures. We speak of two factors, A (temperature) at $a$ levels, B (pressure) at $b$ levels. Each experimental run on the plant is made at some combination of a level of A and a level of B. Such a set of operating conditions is called a treatment combination. We shall be concerned with making runs under all sets of conditions; in this case there are $ab$ treatment combinations. An obvious notation is to represent each treatment combination by two coordinates, the first denoting the level of A, and the second denoting the level of B.

We can carry out a factorial experiment in an incomplete block design by letting the treatment combinations be the "treatments". We shall then wish to subdivide the adjusted treatment sum of squares into components for A, B, AB, ... . If the sums of squares for A, B, AB, ... are independent quadratic forms, the incomplete block design is said in the terminology of J. A. John and Smith (1972) to have factorial structure for that factorial experiment. An equivalent definition under our tacit assumption of normality is that contrasts belonging to different main effects or interactions should be uncorrelated.

The topic of factorial structure has interested several workers. Earlier investigators approached the problem by looking at confounding and partial confounding. Kramer and Bradley (1957) and Zelen (1958) considered the use of group divisible designs for factorial experiments. Later Kurkjian and Zelen (1962, 1963) developed Zelen's ideas further in a special calculus for the analysis of factorial experiments.

Shah (1958, 1960) and Kshirsagar (1966) considered balanced factorial designs. These are designs with factorial structure which have the additional property that for any main effect or interaction all normed contrasts have the same variance. These two authors showed that such designs must belong to the EGD scheme of Hinkelman (1964), which was discussed in Chapter 10.

Pearce (1963) included designs with factorial balance in his classification of designs. His definition was more restrictive than that of Shah and Kshirsagar. We shall use the definition of the latter two authors. J. A. John and Smith (1972) developed sufficient conditions for a design to have factorial structure for a two factor experiment; their criterion was that $\underset{\sim}{N}\underset{\sim}{N}'$ should have a certain cyclic structure, which we shall mention later. Their results were later extended to experiments with three or more factors in Cotter, John, and Smith (1973).

12.2 <u>An example of a $2 \times 3$ factorial experiment</u>. We take the following group divisible design for $t = 6$, which was given in chapter five: 0 1 2, 2 3 4, 4 5 0, 5 1 3. There are three groups: 0 3, 1 4, 2 5; $\lambda_1 = 0$, $\lambda_2 = 1$. Let A denote the first factor with levels 0, 1, and B the second with levels 0, 1, 2. We assign the treatment combinations for the factorial to the treatments of the incomplete block design: $00 \to 0$, $10 \to 3$, $01 \to 1$, $11 \to 4$, $02 \to 2$, $12 \to 5$. The design now becomes

00 01 02; 02 10 11; 11 12 00; 12 01 10.

In writing the concordance matrix we list the treatment combinations by groups in the order:

00 10 01 11 02 12.

Then

$$\underset{\sim}{N}\underset{\sim}{N}' = \begin{bmatrix} r & \lambda_1 & \lambda_2 & \lambda_2 & \lambda_2 & \lambda_2 \\ \lambda_1 & r & \lambda_2 & \lambda_2 & \lambda_2 & \lambda_2 \\ \lambda_2 & \lambda_2 & r & \lambda_1 & \lambda_2 & \lambda_2 \\ \lambda_2 & \lambda_2 & \lambda_1 & r & \lambda_2 & \lambda_2 \\ \lambda_2 & \lambda_2 & \lambda_2 & \lambda_2 & r & \lambda_1 \\ \lambda_2 & \lambda_2 & \lambda_2 & \lambda_2 & \lambda_1 & r \end{bmatrix} .$$

The linear A contrast is

$$-00 + 10 - 01 + 11 - 02 + 12.$$

The vector of coefficients is $\underset{\sim}{c}_1 = (-1, +1, -1, +1, -1, +1)'$. Applying this vector to $\underset{\sim}{N}\underset{\sim}{N}'$ we see that it is a latent vector corresponding to the root $\theta_2 = r - \lambda_1$. The contrast vectors $\underset{\sim}{c}_2 = (-1, -1, 0, 0, +1, +1)' = \text{Lin B}$ and $\underset{\sim}{c}_3 = (1, 1, -2, -2, 1, 1)' = \text{Quad B}$ are latent vectors for the root $\theta_1 = r + \lambda_1 - 2\lambda_2$. The interaction contrasts $\underset{\sim}{c}_4 = (1, -1, 0, 0, -1, 1)' = \text{A} \times \text{Lin B}$ and $\underset{\sim}{c}_5 = (-1, +1, +2, -2, -1, +1)' = \text{A} \times \text{Quad B}$ are latent vectors for the root $\theta_2 = r - \lambda_1$. The contrasts are uncorrelated. For example,

$$\text{cov}(\text{Lin A, Quad B}) = \text{cov}(\underset{\sim}{c}_1'\hat{\underset{\sim}{\tau}}, \underset{\sim}{c}_3'\hat{\underset{\sim}{\tau}}) = \underset{\sim}{c}_1'\underset{\sim}{\Omega}\,\underset{\sim}{c}_3\sigma^2$$

$$= \underset{\sim}{c}_1'\psi_2^{-1}\underset{\sim}{c}_3\sigma^2 = \psi_2^{-1}\underset{\sim}{c}_1'\underset{\sim}{c}_3\sigma^2 = 0, \quad \text{where} \quad \psi_2 = r - \theta_2/k.$$

The key is that the contrasts in which we are interested form a set of mutually orthogonal latent vectors of $\underset{\sim}{N}\underset{\sim}{N}'$, which is a consequence of the selection of a group divisible design.

In computing the sums of squares we note that if $\underset{\sim}{c}_i$ is one of the contrast vectors which is a latent vector of $\underset{\sim}{N}\underset{\sim}{N}'$ with the root $\theta_i$, and if $\psi_i = r - \theta_i/k$, $V(\underset{\sim}{c}_i'\hat{\underset{\sim}{\tau}}) = \underset{\sim}{c}_i'\underset{\sim}{\Omega}\underset{\sim}{c}_i\sigma^2 = \psi_i^{-1}\underset{\sim}{c}_i'\underset{\sim}{c}_i\sigma^2$. The sum of squares for that contrast is $(\underset{\sim}{c}_i'\hat{\underset{\sim}{\tau}})^2\psi_i/(\underset{\sim}{c}_i'\underset{\sim}{c}_i)$. Alternatively, we may write $\underset{\sim}{c}_i\hat{\underset{\sim}{\tau}} = \underset{\sim}{c}_i\underset{\sim}{\Omega}\underset{\sim}{Q} = \underset{\sim}{c}_i\underset{\sim}{Q}/\psi_i$, in which case the sum of squares is $(\underset{\sim}{c}_i'\underset{\sim}{Q})^2/(\psi_i\underset{\sim}{c}_i'\underset{\sim}{c}_i)$. In our example the component sums of squares

A: $(\underset{\sim}{c}_1'\underset{\sim}{Q})^2/8$;

B: $\{3(\underset{\sim}{c}_2'\underset{\sim}{Q})^2 + (\underset{\sim}{c}_3'\underset{\sim}{Q})^2\}/24$;

AB: $\{3(\underset{\sim}{c}_4'\underset{\sim}{Q})^2 + (\underset{\sim}{c}_5'\underset{\sim}{Q})^2\}/12$.

12.3 <u>Factorial structure for the $2^n$ factorials.</u> The contrast vectors for the $2^n$ factorial are, with the addition of the unit vector, the columns of a Hadamard matrix, $\underset{\sim}{H}$. In the case of $2^2$ we have the regression model

$$E(y) = \beta_0 + \beta_1 x_1 + \beta_2 x_2 + \beta_{12} x_1 x_2,$$

where $\beta_0, \beta_1, \beta_2, \beta_{12}$ are the 'effects' and $x_i = \pm 1$. Then

$$\underset{\sim}{\tau} = \begin{bmatrix} (1) \\ a \\ b \\ ab \end{bmatrix} = \begin{bmatrix} 1 & -1 & -1 & +1 \\ 1 & +1 & -1 & -1 \\ 1 & -1 & +1 & -1 \\ 1 & +1 & +1 & +1 \end{bmatrix} \begin{bmatrix} \beta_0 \\ \beta_1 \\ \beta_2 \\ \beta_{12} \end{bmatrix} = \underset{\sim}{H}\underset{\sim}{\beta} .$$

In the general case $\underset{\sim}{H}'\underset{\sim}{H} = \underset{\sim}{H}\underset{\sim}{H}' = 2^n\underset{\sim}{I}$. Without blocking $\hat{\underset{\sim}{\beta}} = \underset{\sim}{H}^{-1}\hat{\underset{\sim}{\tau}} = 2^{-n}\underset{\sim}{H}'\hat{\underset{\sim}{\tau}}$, and $\text{cov}(\hat{\underset{\sim}{\beta}}) = \dfrac{2^{-n}\sigma^2\underset{\sim}{I}}{r}$. With blocking we still have $\hat{\underset{\sim}{\beta}} = 2^{-n}\underset{\sim}{H}'\hat{\underset{\sim}{\tau}}$, but $\text{cov}(\hat{\underset{\sim}{\beta}})$ is different. We now partition the effects vector, $\underset{\sim}{\beta} = (\beta_0, \underset{\sim}{\beta}_1')'$, and the Hadamard matrix, $\underset{\sim}{H} = (\underset{\sim}{1}, \underset{\sim}{H}_1)$; $\underset{\sim}{H}_1'\underset{\sim}{H}_1 = \underset{\sim}{H}_1\underset{\sim}{H}_1' = \underset{\sim}{I}$. Then

$$\hat{\underset{\sim}{\beta}}_1 = 2^{-n}\underset{\sim}{H}_1'\hat{\underset{\sim}{\tau}}, \qquad \text{cov}(\underset{\sim}{\beta}_1) = 2^{-2n}\underset{\sim}{H}_1'\underset{\sim}{\Omega}\underset{\sim}{H}_1\sigma^2.$$

We wish $\text{cov}(\hat{\underset{\sim}{\beta}}_1)$ to be a diagonal matrix. A necessary and sufficient condition for factorial structure is thus that the orthogonal matrix $2^{-n/2}\underset{\sim}{H}$ shall diagonalize $\underset{\sim}{\Omega}$, and, hence, $\underset{\sim}{N}\underset{\sim}{N}'$; equivalently, the usual estimating contrasts for the $2^n$ factorial shall be latent vectors of $\underset{\sim}{N}\underset{\sim}{N}'$.

Let $\underset{\sim}{c}_i$ be the contrast vector for an effect $\beta_i$; let $\theta_i$ be the corresponding latent root of $\underset{\sim}{N}\underset{\sim}{N}'$, and let $\psi_i = r - \theta_i/k$. Then $V(\underset{\sim}{c}_i'\hat{\underset{\sim}{\tau}}) = 2^n\sigma^2\psi_i^{-1}$, and the sum of squares for $\beta_i$ is

$$\hat{\beta}_i^2\psi_i/(\underset{\sim}{c}_i'\underset{\sim}{c}_i) = \hat{\beta}_i^2\psi_i/2^n = (\underset{\sim}{c}_i'\underset{\sim}{Q})^2/(2^n\psi_i).$$

We may now compare the variance of an estimating contrast to the variance in the randomized complete block situation. The efficiency of the estimate $\beta_i$ is thus $\psi_i/r$.

It is not necessary for us to compute $\underset{\sim}{\Omega}$. We need only to compute the latent root, $\psi_i$, that goes with each effect, carry out Yates' algorithm using the $Q_i$ as the 'yields', and divide by $\psi_i$.

The standard example of a balanced confounding scheme for a $2^n$ factorial is that of the $2^3$ carried out in a balanced incomplete block design with $t = 8$, $b = 14$, $r = 7$, $k = 4$, $\lambda = 3$. This is a resolvable design. In each replicate one of the effects A, B, C, AB, AC, BC, ABC is confounded with the two blocks. In the first replicate A is confounded, in the second B is confounded, and so on. One method of analysis is to estimate $\beta_i$ by $\tilde{\beta}_i$, the average of the estimates from the six replicates in which $\beta_i$ is not confounded; $V(\tilde{\beta}_i)$ is thus $\sigma^2/48$.

We can now show that this is indeed the least squares estimate. Apart from $\theta_0 = rk$, the only latent root of $\underset{\sim\sim}{NN'}$ is $\theta_1 = r - \lambda = 4$, corresponding to $\psi = 7 - 1 = 6$. The least squares estimate of $\beta_i$ has $V(\hat{\beta}_i) = \sigma^2/(2^n\psi)$ $= \sigma^2/48 = V(\tilde{\beta}_i)$, and so, since the least squares estimator is the unique minimum variance unbiased estimator, $\tilde{\beta}_i$ and $\hat{\beta}_i$ are identical.

It should be noted that the property of factorial structure in this example does not depend upon the resolvability of the design or upon the balanced partial confounding. What is important is that the design be a balanced incomplete block design. Every one of the four non-isomorphic versions of the BIBD with $t = 8$, $b = 14$, $r = 7$, $k = 4$, $\lambda = 3$, including the three that are not resolvable, has factorial structure for the $2^3$ factorial.

12.4 Group divisible designs for $2^n$ factorials. We introduce this topic by continuing the example of the $2^3$ design in blocks of four plots each. The generalization will then be obvious. Suppose that the replicate in which ABC is confounded between the blocks is omitted. We now have a group divisible design with two groups (1), ab, ac, bc and a, b, c, abc, which correspond to the omitted blocks; we may say that ABC is 'confounded' with the groups. For this design, $r = 6$, $k = 4$, $m = 2$, $n = 4$, $\lambda_1 = 2$, $\lambda_2 = 3$. If, when constructing $\underset{\sim\sim}{NN'}$, we arrange the treatment combinations by groups, we can readily see that the ABC contrast vector is a latent vector corresponding to the simple root $\theta_1 = r + 3\lambda_1 - 4\lambda_2$. The other contrasts are latent vectors for $\theta_2 = r - \lambda_1$.

In this example, $\theta_1 = 0$, $\theta_2 = 4$, $\psi_1 = 6$, $\psi_2 = 5$ and so $V(\hat{\beta}_{123}) = \sigma^2/48$. For the other effects, $V(\hat{\beta}_i) = \sigma^2/40$. This is again equivalent to estimating each effect from those replicates in which it is not confounded.

More generally, we may divide the $2^n$ treatment combinations into $2^p$ groups of size $2^{n-p}$, allocating the treatment combinations to groups as if the groups were the blocks in a pattern for confounding $2^p - 1$ effects. The contrasts for the 'confounded' effects will then be latent vectors for $\theta_1 = rk - 2^n\lambda_2$; the other contrasts are vectors for $\theta_2 = r - \lambda_1$. The relative efficiencies are $E_1 = 2^n\lambda_2/(rk)$ and $E_2 = 1 - (r - \lambda_1)/(rk)$.

Continuing our example, if we retain only the three replicates

| | | | | | | | | | | | |
|---|---|---|---|---|---|---|---|---|---|---|---|
| (1) | ab | c | abc | (1) | ac | b | abc | (1) | bc | a | abc |
| a | b | ac | bc | a | c | ab | bc | b | c | ab | ac |

we have four groups (1) abc; a bc; b ac; ab c. This corresponds to 'confounding' AB, AC, BC with $\lambda_1 = 3$, $\lambda_2 = 1$. Then $V(\hat{\beta}_{12}) = V(\hat{\beta}_{13}) = V(\hat{\beta}_{23}) = \sigma^2/16$; $v(\hat{\beta}_1) = V(\hat{\beta}_2) = V(\hat{\beta}_3) = V(\hat{\beta}_{123}) = \sigma^2/24$; $E_1 = 2/3$, $E_2 = 1$.

12.5 Factors at several levels. When one or more of the factors appears at more than two levels, the corresponding sum of squares involves several contrasts. The necessary and sufficient condition given for $2^n$ designs has to be modified to accommodate this. We shall give the condition for an experiment with two factors; the extension to three or more factors will be obvious.

Suppose that we have two factors, A at a levels and B at b levels. The adjusted sum of squares for treatments has $ab - 1$ degrees of freedom; the corresponding vector subspace (the estimation subspace) has rank $ab - 1$, Scheffé (1959). We wish to divide this into three subspaces for A, B and AB. A necessary and sufficient condition for the design to have factorial structure is that the A, B and AB subspaces should each be spanned by a set of mutually orthogonal latent vectors of $\underset{\sim}{\Omega}$; in the case of the partially balanced designs that we are considering they are also latent vectors of $\underset{\sim}{N}\underset{\sim}{N}'$.

In our example for the $2 \times 3$ factorial, the A and AB subspaces are both spanned by sets of orthogonal latent vectors corresponding to the root $\theta_2 = r - \lambda_1$, while the vectors spanning the B subspace are associated with $\theta_1 = r - (n-1)\lambda_1 + n\lambda_2$.

The reader will see immediately that we can always obtain factorial structure for the two factor experiment by taking a group divisible scheme with the groups of treatment combinations corresponding to the levels of one of the factors. More generally, the rectangular scheme, which is the EGD scheme for two coordinates, and of which the group divisible scheme is a special case, has factorial structure. Contrasts in the A subspace are latent vectors corresponding to $\theta_{01}$; the B subspace is spanned by the latent vectors for $\theta_{10}$ and the AB subspace is spanned by the vectors for $\theta_{11}$.

The designs mentioned above are balanced in the sense of Shah and Kshirsagar. Indeed it is a necessary and sufficient condition for a design to have factorial balance that all the contrasts for A be latent vectors for the same latent root, that the contrasts for B are vectors for the same root, and similarly for AB. A consequence of the property of factorial balance is that, since, for example, all normed A contrasts have the same variance, $\psi^{-1}\sigma^2$, we can speak of the efficiency (of estimation) of A as given by $\psi/r$. The experimenter who has a choice of designs will usually choose a design, if he can find one, with high efficiencies for the main effects, and comparatively low efficiencies, if necessary, for the multifactor interactions. There is, in the following sense, an upper bound on the overall efficiency. Kshirsagar (1958) referred to the loss of efficiency in the estimation of a contrast as $1 - \psi/r$. The total loss of efficiency for a complete set of $t - 1$ orthogonal contrasts is then

$$t - 1 - r^{-1}\mathrm{tr}(\underset{\sim}{C}) = t - 1 - rt(k - 1)/(rk)$$
$$= (t - k)/k.$$

This does not involve the $\lambda_i$; all we can hope to achieve in the way of optimal design for a given block size is to be able to find a set of $\lambda_i$ which will maximize some measure of efficiency subject to this restriction imposed by the block size.

J. A. John and Smith (1972) have approached the problem from a different point of view. Following Kurkjian and Zelen, they have obtained sufficient conditions on $\underset{\sim}{\Omega}$ for a design to have factorial structure.

We again consider two factors, A at a levels and B at b levels. John and Smith write the treatment combinations in ascending order of the levels of one of the factors, say A, and partition $\underset{\sim}{N}\underset{\sim}{N}'$ into submatrices of order a. Their conditions for factorial structure are:

(i) $\underset{\sim}{\Omega}$ has to be a circulant matrix with respect to the submatrices, $\underset{\sim}{\Omega}_i$. The first row of submatrices is $(\underset{\sim}{\Omega}_0, \underset{\sim}{\Omega}_1, \underset{\sim}{\Omega}_2, \ldots, \underset{\sim}{\Omega}_b)$. The second row is $(\underset{\sim}{\Omega}_b, \underset{\sim}{\Omega}_0, \underset{\sim}{\Omega}_1, \ldots, \underset{\sim}{\Omega}_{b-1})$, etc.

(ii) each of the submatrices has the property that its rows sum to the same total and its columns sum to the same total, i.e., $\underset{\sim}{1}$ is a latent vector of $\underset{\sim}{\Omega}_i$ and of $\underset{\sim}{\Omega}_i'$ for each i. These requirements are satisfied by the rectangular designs.

We have shown that, for partially balanced incomplete designs, $\underset{\sim}{N}\underset{\sim}{N}'$ and $\underset{\sim}{\Omega}$ have the same structure, and so the criterion of John and Smith applies equally well to the concordance matrix $\underset{\sim}{N}\underset{\sim}{N}'$. We are thus in the fortunate position with partially balanced designs of being able to tell whether they have factorial structure without matrix inversion. We have already seen in the group divisible examples in Sections 12.2 and 12.4 that, if the treatment combinations can be appropriately assigned in a group divisible scheme, designs with that assignment have factorial structure for all values of $\lambda_i$.

The criterion is thus more broadly applicable to association schemes rather than to individual designs.

There are designs with factorial structure which do not have factorial balance. We give an example for the $3^2$ factorial which will also illustrate the criterion of John and Smith. It involves the use of the LS3 association scheme for nine treatments. The LS3 scheme has three associate classes. The treatments are arranged in a square array upon which a latin square is superimposed. Two treatments are first associates if they occur in the same row or in the same column, second associates if they share the same letter, and third associates otherwise; if $\lambda_2 = \lambda_3$ the scheme reduces to the $L_2$ scheme. For our example we take the following array and square:

$$\begin{matrix} 00 & 01 & 02 \\ 10 & 11 & 12 \\ 20 & 21 & 22 \end{matrix}, \qquad \begin{matrix} \alpha & \beta & \gamma \\ \gamma & \alpha & \beta \\ \beta & \gamma & \alpha \end{matrix}.$$

Then

$$\underset{\sim\sim}{NN'} = \begin{bmatrix} \underset{\sim}{U}_0 & \underset{\sim}{U}_1 & \underset{\sim}{U}_2 \\ \underset{\sim}{U}_2 & \underset{\sim}{U}_0 & \underset{\sim}{U}_1 \\ \underset{\sim}{U}_1 & \underset{\sim}{U}_2 & \underset{\sim}{U}_0 \end{bmatrix},$$

where

$$\underset{\sim}{U}_0 = \begin{bmatrix} r & \lambda_1 & \lambda_1 \\ \lambda_1 & r & \lambda_1 \\ \lambda_1 & \lambda_1 & r \end{bmatrix}, \quad \underset{\sim}{U}_1 = \begin{bmatrix} \lambda_1 & \lambda_2 & \lambda_3 \\ \lambda_3 & \lambda_1 & \lambda_2 \\ \lambda_2 & \lambda_3 & \lambda_1 \end{bmatrix}, \quad \underset{\sim}{U}_2 = \begin{bmatrix} \lambda_1 & \lambda_3 & \lambda_2 \\ \lambda_2 & \lambda_1 & \lambda_3 \\ \lambda_3 & \lambda_2 & \lambda_1 \end{bmatrix}.$$

In constructing $\underset{\sim\sim}{NN'}$ we have taken the treatment combinations in the order 00 01 02 10 11 12 20 21 22, and have let the first coordinate correspond to the level of A and the second to the level of B. The latent roots of $\underset{\sim\sim}{NN'}$, in addition to rk, are

$$\begin{aligned} \theta_1 &= r + \lambda_1 - \lambda_2 - \lambda_3, & \alpha_1 &= 4, \\ \theta_2 &= r - 2\lambda_1 + 2\lambda_2 - \lambda_3, & \alpha_2 &= 2, \\ \theta_3 &= r - 2\lambda_1 - \lambda_2 + 2\lambda_3, & \alpha_3 &= 2. \end{aligned}$$

The contrasts for A and B are latent vectors for $\theta_1$. The contrast Lin A Lin B is not a latent vector of $\underset{\sim\sim}{NN'}$. However in this case the AB subspace splits into two portions corresponding to the partitioning of the interaction into what Bose and Yates called the AB and $AB^2$ components or the I and J components. These correspond to hyperplanes.

The hyperplanes $x_1 + x_2 \equiv 0, 1, 2, \bmod 3$, each contain three points: 00, 12, 21 on $x_1 + x_2 \equiv 0$; 01, 10, 22 on $x_1 + x_2 \equiv 1$; 02, 11, 20 on $x_1 + x_2 \equiv 2$. The vector $\underset{\sim}{c}_1 = (1, 0, -1, 0, -1, 1, -1, 1, 0)'$ corresponds to $00 + 12 + 21 - 02 - 11 - 20$, or $\{x_1 + x_2 \equiv 0\} - \{x_1 + x_2 \equiv 2\}$. The vector $\underset{\sim}{c}_2 = (1, -2, 1, -2, 1, 1, 1, 1, -2)'$ corresponds to $\{x_1 + x_2 \equiv 0\} - 2\{x_1 + x_2 \equiv 1\} + \{x_1 + x_2 \equiv 2\}$. Each of them is a latent vector for $\theta_2$. Similarly, contrasts between the subsets $\{x_1 + 2x_2 \equiv 0\}(00, 11, 22)$, $\{x_1 + 2x_2 \equiv 1\}(02, 10, 21)$ and $\{x_1 + 2x_2 \equiv 2\}(01, 12, 20)$ give rise to latent vectors $\underset{\sim}{c}_3 = (1, -1, 0, 0, 1, -1, -1, 0, 1)'$ and $\underset{\sim}{c}_4 = (1, 1, -2, -2, 1, 1, 1, -2, 1)'$ for $\theta_3$.

We may consider the AB interaction subspace as being further subdivided into $(x_1 + x_2)$ and $(x_1 + 2x_2)$ subspaces; $\underset{\sim}{c}_1$ and $\underset{\sim}{c}_2$ span the $(x_1 + x_2)$ subspace, while $\underset{\sim}{c}_3$ and $\underset{\sim}{c}_4$ span the $(x_1 + 2x_2)$ subspace. The Lin A Lin B vector, $\underset{\sim}{c}_5 = (1, 0, -1, 0, 0, 0, -1, 0, 1)'$ is in neither subspace. However, we have the relationship $6\underset{\sim}{c}_5 = 3\underset{\sim}{c}_1 - \underset{\sim}{c}_2 + 3\underset{\sim}{c}_3 + \underset{\sim}{c}_4$, and $\underset{\sim}{c}_5$ is clearly orthogonal to any contrasts in the main effects for A or B.

12.6 <u>Several factors</u>. The criterion that the subspaces be spanned by sets of mutually orthogonal latent vectors of $\underset{\sim}{\Omega}$ still holds when there are several factors. It can be shown that a sufficient (but again not necessary) condition for a design to have factorial structure for n factors is that it be partially balanced with the EGD scheme for n coordinates. With three factors the main effects of A, B, C correspond to the roots $\theta_{100}$, $\theta_{010}$, $\theta_{001}$ of $\underset{\sim\sim}{NN'}$ respectively; AB, AC and BC correspond to $\theta_{110}$, $\theta_{101}$, $\theta_{011}$ and ABC to $\theta_{111}$.

Cotter, John and Smith (1973) extended the results of John and Smith (1972) to several factors. Their sufficient condition calls for a pseudocyclic structure in submatrices $\underset{\sim}{\Omega}_{ij}$, which have the property that $\underset{\sim}{1}$ is a latent vector of both $\underset{\sim}{\Omega}_{ij}$ and $\underset{\sim}{\Omega}_{ji}$. For example, with three factors A, B, C at a, b and c levels, they ask first that $\underset{\sim}{\Omega}$ be partitioned into submatrices of order bc which satisfy their criterion for two factors. This ensures that A is independent of B and C. They then partition each of the matrices $\underset{\sim}{\Omega}_i$ into submatrices $\underset{\sim}{\Omega}_{ij}$ of order c which satisfy the criterion for two factors, thus establishing the independence of B and C.

12.7 <u>The cubic scheme for the $3^3$ factorial</u>. The cubic scheme for $t = 27$ has factorial structure for the $3^3$ factorial experiment. The scheme is a degenerate version of the EGD scheme for $n = 3$. The associate classes are combined in the following way: the first associates in the cubic scheme are the 001, 010, 100 associates in the EGD scheme; the second associates correspond to 011, 101, 110 and the third associates to 111. The six degrees of freedom for the main effects of A, B, C are, therefore, associated with $\theta_1$; the two factor interactions are associated with $\theta_2$ and the three factor interaction with $\theta_3$.

The traditional partially confounded designs can be obtained from the scheme. For example, when the three initial blocks 000, 001, 002; 000, 010, 020; 000, 100, 200 are developed, we get a total of eighty-one blocks, but only twenty-seven of them are distinct. The twenty-seven distinct blocks form a resolvable design with $t = b = 27$, $r = k = 3$, $\lambda_1 = 1$, $\lambda_2 = \lambda_3 = 0$. In one replicate the main effects of A and B and the AB interaction are confounded with blocks. In another replicate A, C and AC are confounded, and in the third B, C and BC. Further designs in which $k$ is a multiple of three, and in which the blocks are obtained as hyperplanes or as combinations of hyperplanes, are found in the standard references. The following designs have values of $k$ which are not multiples of three, and cannot be obtained by the traditional partial confounding techniques.

The initial block 000, 011, 101, 110 gives a design with $t = b = 27$, $r = k = 4$, $\lambda_2 = 1$, $\lambda_1 = \lambda_3 = 0$, $\theta_1 = 4$, $\theta_2 = 1$, $\theta_3 = 7$. The relative efficiencies are thus $E_1 = 0.75$ for the main effects, $E_2 = 15/16$ for the two factor interactions, and $E_3 = 9/16$ for ABC. (We recall that $E_i = \psi_i/r$).

The initial block 000, 111, 022, 202, 220 gives a symmetric design with $\lambda_1 = 0$, $\lambda_2 = \lambda_3 = 1$. This is perhaps more attractive to some experimenters inasmuch as $E_1 > E_2$; $E_1 = 0.96$, $E_2 = 0.84$, $E_3 = 0.72$.

The E'S(1) design with initial block 000, 001, 002, 010, 020, 100, 200 has $t = b = 27$, $r = k = 7$, $\lambda_1 = 3$, $\lambda_2 = 2$, $\lambda_3 = 0$, $E_1 = 0.67$, $E_2 = 0.98$, $E_3 = 0.92$. This design is not particularly attractive for estimating main effects.

The ES(3) design in which the $j^{th}$ block consists of the third associates of the $j^{th}$ treatment has $t = b = 27$, $r = k = 8$, $\lambda_1 = 4$, $\lambda_2 = 2$, $\lambda_3 = 1$, $E_1 = 0.75$, $E_2 = 0.94$, $E_3 = 0.98$.

12.8 Designs for the $2^4$ factorial. It is well known that the latin square, $L_2$, scheme has factorial structure for the $2^4$ experiment. We take the basic array

| | | | |
|---|---|---|---|
| 1 | a | b | ab |
| c | ac | bc | abc |
| d | ad | bd | abd |
| cd | acd | bcd | abcd . |

The A, B, AB, C, D, CD contrasts are latent vectors for $\theta_1 = r + 2\lambda_1 - 3\lambda_2$; the other contrasts are latent vectors for $\theta_2 = r - 2\lambda_1 + \lambda_2$. The concordance matrix may be written as

$$\underset{\sim}{N}\underset{\sim}{N}' = \begin{bmatrix} \underset{\sim}{A} & \underset{\sim}{B} & \underset{\sim}{B} & \underset{\sim}{B} \\ \underset{\sim}{B} & \underset{\sim}{A} & \underset{\sim}{B} & \underset{\sim}{B} \\ \underset{\sim}{B} & \underset{\sim}{B} & \underset{\sim}{A} & \underset{\sim}{B} \\ \underset{\sim}{B} & \underset{\sim}{B} & \underset{\sim}{B} & \underset{\sim}{A} \end{bmatrix}$$

where $\underset{\sim}{A} = (r - \lambda_1)\underset{\sim}{I}_4 + \lambda_1 \underset{\sim}{J}_4$, $\underset{\sim}{B} = (\lambda_1 - \lambda_2)\underset{\sim}{I} + \lambda_2 \underset{\sim}{J}$. The criterion of Cotter, John and Smith is clearly satisfied.

We mentioned in Chapter 8 that there are two nonisomorphic latin squares of side four. They lead to two $L_3$ schemes which we called $L_3(4)$ and $L_3^*(4)$. The reader can show by writing down the concordance matrices that the $L_3$ scheme formed by using the $L_3(4)$ square has factorial structure, while the other scheme does not. It will be recalled that changing the first and second associates in the $L_3^*(4)$ scheme leads to the pseudo $L_2$ scheme. The pseudo $L_2$ scheme does not therefore have factorial structure.

We can also extend the idea of the cubic scheme to the hypercubic association scheme for $2^4$ treatments. We represent each treatment combination by four coordinates $x_1x_2x_3x_4$. In this case $x_i = 0$ or $1$ and arithmetic is mod 2. Two treatments are $i^{th}$ associates if their representations disagree in exactly $i$ coordinates. The reader may derive the $\underset{\sim}{P}$ matrices for himself. The latent roots of $\underset{\sim}{N}\underset{\sim}{N}'$ and their multiplicities are

$$\begin{aligned} \theta_1 &= r + 2\lambda_1 - 2\lambda_3 - \lambda_4, & \alpha_1 &= 4, \\ \theta_2 &= r - 2\lambda_2 + \lambda_4, & \alpha_2 &= 6, \\ \theta_3 &= r - 2\lambda_1 + 2\lambda_3 - \lambda_4, & \alpha_3 &= 4, \\ \theta_4 &= r - 4\lambda_1 + 6\lambda_2 - 4\lambda_3 + \lambda_4, & \alpha_4 &= 1. \end{aligned}$$

The A, B, C and D contrasts are latent vectors for $\theta_1$, the two factor interactions give vectors for $\theta_2$, three factor interactions for $\theta_3$ and the ABCD contrast is the vector for $\theta_4$.

The scheme degenerates when some of the $\lambda_i$ are equal. If $\lambda_1 = \lambda_3$, $\theta_1 = \theta_3$. The scheme is then hierarchic group divisible with two groups of eight treatments, each divided into four subgroups. The groups are:

0000, 1111; 0011, 1100; 0101, 1010; 0110, 1001 and
0001, 1110; 0010, 1101; 0100, 1011; 1000, 0111.

If either (i) $\lambda_1 = \lambda_2 = \lambda_1'$ and $\lambda_3 = \lambda_4 = \lambda_2'$ or (ii) $\lambda_1 = \lambda_4 = \lambda_2'$ and $\lambda_2 = \lambda_3 = \lambda_1'$, the degenerate scheme is the negative latin square scheme of Mesner. This is the scheme obtained by letting $n = -4$ and $i = -2$ in the $\underset{\sim}{P}$ matrices for the $L_2$ scheme. It has $n_1 = 10$ and $n_2 = 5$;

$$\underset{\sim}{P}_1 = \begin{pmatrix} 6 & 3 \\ 3 & 2 \end{pmatrix}, \qquad \underset{\sim}{P}_2 = \begin{pmatrix} 6 & 4 \\ 4 & 0 \end{pmatrix}.$$

The latent roots of $\underset{\sim}{N}\underset{\sim}{N}'$ are

$$\theta_1 = r + \lambda_1' - 2\lambda_2', \qquad \alpha_1 = 10,$$
$$\theta_2 = r - 3\lambda_1' + 2\lambda_2', \qquad \alpha_2 = 5.$$

There are two designs with this scheme which have $k = 5$. They are isomorphic. The E'S(1) design has $\lambda_1 = \lambda_2 = 2$, $\lambda_3 = \lambda_4 = 0$. In this design the main effects and the four factor interaction are associated with the same root; E(A) = E(ABCD) = 0.64, E(AB) = E(ABC) = 0.96. In the E'S(3) design the situation is reversed; $\lambda_1 = \lambda_4 = 0$, $\lambda_2 = \lambda_3 = 2$; E(A) = E(AB) = 0.96, E(ABC) = E(ABCD) = 0.64.

# REFERENCES


Bhattacharya, K. N. (1944), On a new symmetrical balanced incomplete block design. Bull. Calcutta Math. Soc., **36**, 91-96.

Bose, R. C. (1938), On the application of the properties of Galois fields to the problem of the construction of hyper-graeco-latin squares. Sankhya, **6**, 105-110.

Bose, R. C. (1939), On the construction of balanced incomplete block designs. Annals of Eugenics, **9**, 353-399.

Bose, R. C. (1949), A note on Fisher's inequality for balanced incomplete block designs. Ann. Math. Statist., **20**, 619-620.

Bose, R. C. (1963), Combinatorial properties of partially balanced designs and association schemes. Sankhya, A, **25**, 109-136.

Bose, R. C. and Connor, W. S. (1952), Combinatorial properties of group divisible incomplete block designs. Ann. Math. Statist., **23**, 367-383.

Bose, R. C., Chatworthy, W. H. and Shrikhande, S. S. (1954), Tables of partially balanced designs with two associate classes. North Carolina Agric. Exp. Station Tech. Bull. No. 107.

Bose, R. C. and Mesner, D. M. (1959), On linear associative algebras corresponding to association schemes of partially balanced designs. Ann. Math. Statist., **30**, 21-38.

Bose, R. C. and Nair, K. R. (1939), Partially balanced incomplete block designs. Sankhya, **4**, 337-372.

Bose, R. C. and Shimamoto, T. (1952), Classification and analysis of designs with two associate classes. J. Amer. Statist. Assoc., **47**, 151-190.

Bose, R. C., Shrikhande, S. S. and Bhattacharya, K. N. (1953), On the construction of group divisible incomplete block designs. Ann. Math. Statist., **24**, 167-195.

Bruck, R. H. and Ryser, H. J. (1949), The non-existence of certain finite projective planes. Canad. J. Math., **1**, 88-93.

Chang, L. C. (1960), Association schemes of partially balanced designs with parameters $v = 28$, $n_1 = 12$, $n_2 = 15$, $p^2_{11} = 4$. Science Record, **4**, new series, 12-18.

Chowla, S. and Ryser, H. J. (1950), Combinatorial problems. Canad. J. Math., **2**, 93-99.

Clatworthy, W. H. (1956), Contributions on partially balanced incomplete block designs with two associate classes. National Bureau of Standards Appl. Math. Series, 47.

Clatworthy, W. H. (1973), Tables of two-associate-class partially balanced designs. National Bureau of Standards Appl. Math. Series, 63.

Connor, W. S. (1958), The uniqueness of the triangular association scheme. Ann. Math. Statist., **29**, 262-266.

Cotter, S. C., John, J. A. and Smith, T. M. F. (1973), Multi-factor experiments in non-orthogonal designs. J. R. Statist. Soc., B, **35**, 361-367.

Fisher, R. A. (1940), An examination of the different possible solutions of a problem in incomplete blocks. Annals of Eugenics, **10**, 52-75.

Foody, W. and Hedayat, A. (1977), On theory and applications of BIB designs with repeated blocks. Ann. Statist., **5**, 932-945.

Hall, M. and Connor, W. S. (1953), An embedding theorem for incomplete block designs. Canad. J. Math., **6**, 35-41.

Hanani, H. (1961), The existence and construction of balanced incomplete block designs. Ann. Math. Statist., **32**, 361-386.

Hanani, H. (1965), A balanced incomplete block design. Ann. Math. Statist., **36**, 711.

Hedayat, A. and John, P. W. M. (1974), Resistant and susceptible BIB designs. Ann. Statist., **2**, 148-158.

Hinkelmann, K. (1964), Extended group divisible partially balanced incomplete block designs. Ann. Math. Statist., **35**, 681-695.

Hinkelmann, K. and Kempthorne, O. (1963), Two classes of group divisible partial diallel crosses. Biometrika, **50**, 281-291.

Hoffman, A. J. (1960), On the uniqueness of the triangular association scheme. Ann. Math. Statist., **31**, 492-497.

James, A. T. and Wilkinson, G. N. (1971), Factorization of the residual operator and canonical decomposition of nonorthogonal factors in the analysis of variance. Biometrika, **58**, 279-294.

Jarrett, Richard G. (1977), Bounds for the efficiency factor of block designs. Biometrika, **64**, 67-72.

John, J. A. (1966), Cyclic incomplete block designs. J. R. Statist. Soc., B, **28**, 345-360.

John, J. A. and Smith, T. M. F. (1972), Two-factor experiments in non-orthogonal designs. J. R. Statist. Soc., B, **34**, 401-409.

John, J. A., Wolock, F. and David, H. A. (1972), Cyclic Designs, National Bureau of Standards (U.S.) Appl. Math. Series, 62.

John, P. W. M. (1964), Balanced designs with unequal numbers of replicates. Ann. Math. Statist., **35**, 897-899.

John, P. W. M. (1970), The non-existence of linked block designs with Latin square association schemes. Ann. Math. Statist., **41**, 1105-1107.

John, P. W. M. (1971), Statistical Design and Analysis of Experiments. Macmillan, New York.

John, P. W. M. (1973), A balanced design for eighteen varieties. Technometrics, **15**, 641-642.

John, P. W. M. (1976), Robustness of balanced incomplete block designs. Ann. Statist., **4**, 960-962.

John, P. W. M. (1978a), A note on concordance matrices for partially balanced designs. Utilitas Mathematica, **14**, 3-8.

John, P. W. M. (1978b), A note on a finite population plan. Ann. Statist., **6**, 697-699.

Kageyama, S. (1962), On the reduction of associate classes for certain PBIB designs. Ann. Math. Statist., **43**, 1528-1540.

Kempthorne, O. (1953), A class of experimental designs using blocks of two plots. Ann. Math. Statist., **24**, 76-84.

Kempthorne, O. (1956), The efficiency factor of an incomplete block design. Ann. Math. Statist., **27**, 846-849.

Kiefer, J. (1959), Optimum experimental designs. J. R. Statist. Soc., B, **21**, 272-319.

Kramer, C. Y. and Bradley, R. A. (1957), Intrablock analysis for factorials in two associate class group divisible designs. Ann. Math. Statist., **28**, 349-361.

Kshirsagar, A. (1958), A note on the total relative loss of information in any design. Calcutta Stat. Assn. Bull., **7**, 78-81.

Kshirsagar, A. (1966), Balanced factorial designs. J. R. Statist. Soc., B, **28**, 559-567.

Kurkjian, B. and Zelen, M. (1962), A calculus for factorial arrangements. Ann. Math. Statist., **33**, 600-619.

Kurkjian, B. and Zelen, M. (1963), Applications of the calculus of factorial arrangements I. Block and direct product designs. Biometrika, **50**, 63-73.

Mesner, D. M. (1965), A note on the parameters of PBIB association schemes. Ann. Math. Statist., 36, 331-336.

Mesner, D. M. (1967), A new family of partially balanced incomplete block designs with some latin square properties. Ann. Math. Statist., 38, 571-581.

Patterson, H. D., and Williams, E. R. (1976), Some theoretical results on general block designs. Proc. 5th British Combinatorial Conference, Congressus Numerantium xv, eds. C.St.J. Nash Williams and J. Schekan, 489-496. Utilitas Mathematica, Winnipeg.

Pearce, S. C. (1963), The use and classification of non-orthogonal designs. J. R. Statist. Soc., A, 126, 353-377.

Pearce, S. C. (1968), The mean efficiency of equi-replicate designs. Biometrika, 55, 251-253.

Pearce, S. C. (1970), The efficiency of block designs in general. Biometrika, 57, 339-346.

Raghavarao, D. (1960), A generalization of group divisible designs. Ann. Math. Statist., 31, 756-765.

Raghavarao, D. (1971), Constructions and Combinatorial Problems in Design of Experiments. Wiley, New York.

Raghavarao, D. and Chandrasekhararao, K. (1964), Cubic designs. Ann. Math. Statist., 35, 389-397.

Rao, C. R. (1946), Difference sets and combinatorial arrangements derivable from finite geometrics. Proc. Natl. Inst. Sci., 12, 123-135.

Rao, C. R. (1947), General methods of analysis for incomplete block designs. J. Amer. Statist. Assoc., 42, 541-561.

Roy, J. (1958), On the efficiency factor of block designs. Sankhya, 19, 181-188.

Roy, P. M. (1953-54), Hierarchical group divisible incomplete block designs with m associate classes. Science and Culture, 19, 210-211.

Scheffé, H. (1959), The Analysis of Variance. Wiley, New York.

Schutzenberger, M. P. (1949), A non-existence theorem for an infinite family of incomplete block designs. Annals of Eugenics, 14, 286-287.

Seiden, E. (1966), A note on the construction of partially balanced incomplete block designs with $v = 28$, $n_1 = 12$, $n_2 = 15$ and $p_{11}^2 = 4$. Ann. Math. Statist., 37, 1783-1789.

Shah, B. V. (1958), On balancing in factorial experiments. Ann. Math. Statist., 29, 776-779.

Shah, B. V. (1960), Balanced factorial experiments. Ann Math. Statist. 31, 502-514.

Shrikhande, S. S. (1950). The impossibility of certain symmetrical balanced incomplete block designs. Ann. Math. Statist., **21**, 106-111.

Shrikhande, S. S. (1952), On the dual of some balanced incomplete block designs. Biometrics, **8**, 66-72.

Shrikhande, S. S. (1959a), On a characterization of the triangular association scheme. Ann. Math. Statist., **30**, 39-47.

Shrikhande, S. S. (1959b), The uniqueness of the $L_2$ association scheme. Ann. Math. Statist., **30**, 781-798.

Shrikhande, S. S. and Singh, N. K. (1962), On a method of constructing symmetrical balanced incomplete block designs. Sankhya, A, **24**, 25-32.

Singla, S. L. (1977), An extension of $L_2$ designs. Austral. J. Statist., **19**, 126-131.

Tocher, K. D. (1952), The design and analysis of block experiments. J. R. Statist. Soc., B, **14**, 45-100.

Vartak, M. N. (1955), On the application of Kronecker product of matrices to statistical designs. Ann. Math. Statist., **26**, 420-438.

Williams, E. R. (1975), Efficiency balanced designs. Biometrika, **62**, 686-687.

Wynn, H. (1977), Convex sets of finite population plans. Ann. Statist., **5**, 414-418.

Yates, F. (1936), Incomplete randomized blocks. Annals of Eugenics, **7**, 121-140.

Zelen, M. (1958), The use of group divisible designs for confounded asymmetrical factorial arrangements. Ann. Math. Statist., **29**, 22-40.